# Study Guide

# The Essentials of Statistics
## A Tool for Social Research

### SECOND EDITION

**Joseph F. Healey**
Christopher Newport University

Prepared by

**Dawn M. Baunach**
Georgia State University

 **WADSWORTH**
CENGAGE Learning™

Australia • Brazil • Japan • Korea • Mexico • Singapore • Spain • United Kingdom • United States

ISBN-13: 978-0-495-60147-0
ISBN-10: 0-495-60147-0

**Wadsworth**
10 Davis Drive
Belmont, CA 94002-3098
USA

Cengage Learning is a leading provider of customized learning solutions with office locations around the globe, including Singapore, the United Kingdom, Australia, Mexico, Brazil, and Japan. Locate your local office at: **www.cengage.com/international**

Cengage Learning products are represented in Canada by Nelson Education, Ltd.

To learn more about Wadsworth, visit
**www.cengage.com/wadsworth**

Purchase any of our products at your local college store or at our preferred online store
**www.ichapters.com**

Printed in the United States of America
1 2 3 4 5 6 7 8 12 11 10 09

# TABLE OF CONTENTS

# PREFACE

This text is designed as a supplement to Healey, <u>The Essentials of Statistics: A Tool for Social Research</u>, Second Edition, 2009. It is not designed as a replacement of your text. Instead, this text provides chapter summaries, lists of key terms, multiple choice questions, additional work problems, and detailed answers for each chapter. Additionally, SPSS and Microcase computer problems apply the statistics from each chapter.

This study guide was written to maximize your understanding of the concepts in your main textbook. After you have finished studying for an exam, try the multiple choice problems at the end of each chapter. Check your answers with those at the end of the chapter. These questions will measure your understanding of the general concepts of each chapter.

Next, try your hand at the work problems. These problems will give you additional practice with data and word problems. Detailed answers to all of the problems are at the end of each chapter. Remember, the more practice you get, the easier it will be and the better you will do.

Finally, if you're learning to use one of the many available computer packages for statistics, try the computer problems. These problems can be solved using SPSS or Microcase. Your main textbook gives easy to use instructions for the computer at the end of each chapter. Reading and interpreting the output will provide you with invaluable hands-on training, not to mention a new skill for your resume.

Many of the examples in this text come from contemporary social problems, but all of the data are fictional. This text is written to be easy to read and easy to understand. Statistics has a bad reputation, but if you sit back and relax a bit, you might find you enjoy it!

# COMPONENTS OF THE STUDY GUIDE

Each chapter of this study guide is divided into the following sections:

- **Learning Objectives:** Each chapter begins with a description of what you should learn by the end of each chapter.

- **Key Terms:** This section lists the key terms for each chapter and the page numbers in your textbook where you can find a definition.

- **Chapter Summary:** The Chapter Summary is a short description of the material covered in that chapter.

- **Chapter Outline:** In the Chapter Outline you will find a detailed presentation of the material covered in that chapter.

- **Multiple Choice Questions:** Twenty multiple choice questions on the chapter material are presented in this section, along with an indication of which Learning Objective they address.

- **Work Problems:** Try your hand at completing the work problems (most of which require hand calculations) to get the most out of the Study Guide. The Work Problems are also accompanied by an indication of which Learning Objective they address.

- **SPSS or Microcase Problems:** Being able to use a computer statistical package is a very useful skill; complete these problems for additional practice review.

- **Answer Key:** The Answer Key includes answers to the Multiple Choice questions and detailed answers to the Work Problems and the SPSS/Microcase Problems (including example computer output). You will also find an indication of where in your textbook you can find the information to answer each Multiple Choice question.

# CHAPTER ONE
## Introduction

*Learning Objectives: By the end of this chapter, you will be able to:*

1. *Describe the limited but crucial role of statistics in social research.*
2. *Distinguish among three applications of statistics (univariate descriptive, bivariate descriptive, and inferential) and identify situations in which each is appropriate.*
3. *Identify and describe three levels of measurement and cite examples of variables from each.*

## KEY TERMS

- ✓ Data  (p. 9)
- ✓ Data reduction  (p. 15)
- ✓ Dependent variable  (p. 11)
- ✓ Descriptive statistics  (p. 15)
- ✓ Hypothesis  (p. 12)
- ✓ Independent variable  (p. 11)
- ✓ Inferential statistics  (p. 16)
- ✓ Level of measurement  (p. 17)
- ✓ Measures of association  (p. 15)
- ✓ Population  (p. 16)
- ✓ Research  (p. 9)
- ✓ Sample  (p. 16)
- ✓ Statistics  (p. 9)
- ✓ Theory  (p. 10)
- ✓ Variable  (p. 11)

## CHAPTER SUMMARY

After you finish this course, you will not be expected to be fluent in statistics.  Instead, you should have a better understanding of the structure of statistics, which tests are appropriate to use when, and which tests to use with which variables.  Some of you will struggle with the course, but in the end you will have a better understanding of research methodologies and statistical tests, and you will be ready to venture on to a more advanced level.

It is important to remember that statistics is a language all its own, with its own properties and slang.  Remind yourself of this fact often, especially when you are having trouble with a phrase or word ordering; you are learning a new language.

Also, keep in mind that statistics is <u>not</u> mathematics. We use mathematical operations to compute the statistics, but the statistics themselves are tools for describing and defining populations of <u>people</u>. I emphasize the word <u>people</u> because that is what we are most often studying: people, groups, societies. For example, we might be interested in the average salary for college graduates or the average GPA for college athletes.

If I could pick only one concept that I feel to be the most important in this chapter, it would be the level of measurement of a variable. If you can look at a variable and know its level of measurement, then you can figure out what statistical techniques can be used to study that variable. As you will see later, there are different statistics for different levels of measurement. Picking the wrong one not only will give you incomplete information, but in many cases it will give you the wrong information.

Another important point to keep in mind about statistics is its role in research. Statistics is a powerful and indispensable tool for testing theory and evaluating hypotheses.

## CHAPTER OUTLINE

### 1.1 Why Study Statistics?
- Statistics are used by social scientists to test their ideas and theories through research, where research involves the gathering of information to test these ideas and theories. While research can take many forms, your text book focuses attention on quantitative research, the kind of research that uses and studies numerical information, called data.

### 1.2 The Role of Statistics in Scientific Inquiry.
- Statistics are one method used to test theory. A theory is an explanation of the relationships between social phenomena, and a hypothesis is a statement about the relationship between variables.
- Hypotheses are often stated in terms of a causal relationship between an independent variable and a dependent variable. The independent variable is the "cause," and the dependent variable is the "effect." For example, human capital theory asserts that people develop their personal "capital" in order to improve their personal economics. Personal, or human, "capital" includes education and occupational training. Using human capital theory, then, we can hypothesize that increases in educational attainment produce increases in income. Theories are more general than hypotheses, which tend to be more specific; theories are usually stated in terms of general concepts, where hypotheses are usually stated in terms of variables. In our hypothesis, educational attainment would be the independent variable, the "cause," and income would be the dependent variable, the "effect."
- A variable is any trait that can change value from case to case, such as race: I may be white, my neighbor may be black. The variable "race" can change from one case (mine) to the next (my neighbor).

## 1.3   The Goals of This Text.

- Statistics are tools used by social scientists to increase our knowledge of the social world.
- A knowledge of and familiarity with statistics provides a basis for more advanced study of social phenomena and will prove very useful in future careers and education.

## 1.4   Descriptive and Inferential Statistics.

- There are two types of statistics which make up the majority of your text book. The first type, descriptive statistics, describes groups on some variable.
  - Your text also distinguishes between two types of descriptive statistics: univariate descriptive and bivariate (or multivariate) descriptive. Univariate descriptive statistics summarize or describe a single variable; bivariate (or multivariate) descriptive statistics describe the relationship(s) between two (or more than two) variables.
  - Descriptive statistics fulfill two useful functions. First, through the process of data reduction, descriptive statistics permit us to use a few numbers to represent many numbers. Also, by using measures of association, researchers can quantify the kind of relationship between social phenomena, with particular emphasis on issues of causation and prediction.
- The second type of statistics, inferential statistics, allows us to <u>infer</u> our results from a sample to the rest of the population.
  - The total collection of all cases in which the researcher is interested is called the population. Perhaps you want to study females of childbearing age in the U.S. Your population then is <u>every</u> female between the ages of 13 and 45 in the United States.
  - A sample is a carefully chosen subset of a defined population. It is virtually impossible to ask every single person in a population their sex, weight, height, or whatever else you are interested in studying, therefore you need a <u>sample</u>, or a subset, of the population.

## 1.5   Level of Measurement.

- Variables are the items on a questionnaire or survey which we are using to measure our concepts. Variables are categorized into three types or "levels" of measurement: nominal, ordinal, and interval-ratio. The level of measurement of a variable refers to the mathematical nature of the variable. Most importantly, a variable's level of measurement tells us which statistical techniques are appropriate.
  - Nominal variables simply classify cases into categories, such as male or female. There is no ranking or ordering of the categories; you cannot say that males are higher than females, or that Protestants are higher than non-Protestants. Categories of nominal variables can only be compared in terms of their relative size; no other mathematical operations (addition, subtraction, etc.) are permissible. Examples of nominal variables are: gender, religious preference, race/ethnicity, all Yes/No questions.

- The categories for nominal variables must be *mutually exclusive* (each subject fits into one and only one category), *exhaustive* (there must be a category which every subject can fall in to), and *homogeneous* (the categories must be similar enough to allow generalization). Healy gives the example of scales for religious categories. In the first scale, a person who is Episcopalian could fall into more than one category: Episcopalian and Protestant. Therefore, that categorization is not mutually exclusive. In the second scale, a person who is Muslim would not fall into any of the categories, therefore the scale is not exhaustive. In the third scale, those people who would fall into the category "Non-Protestant" (which would include Catholics, Muslims, Buddhists, Others, or None) are so dissimilar with respect to religious beliefs, that this categorization is not homogeneous. Only the fourth scale meets all three of these criteria.
- Ordinal variables have the same qualities of nominal variables, but go one step further -- the categories can be <u>ranked</u> or <u>ordered</u> from high to low, such as class standing: are you a freshman, sophomore, junior, or senior? Not only can you be classified into one and only one category, but we can <u>rank-order</u> these categories based on seniors having more education and/or unit hours than juniors, etc. Note that there is no true zero point, meaning that we cannot say that being a senior is three times higher than being a freshman, only that being a senior <u>is</u> higher than being a freshman. Examples of ordinal variables are: class ranking, social class, most scales (especially those that are measured from "strongly agree" to "strongly disagree").
- Interval-ratio variables go even a step further than ordinal variables. These variables are classified into categories (as are nominal variables) and are rank-ordered (as are ordinal variables), but most importantly, the <u>exact</u> <u>distance</u> between the categories is <u>known</u>. For example, if you were to classify all of your classmates as freshmen, sophomores, juniors, or seniors, we would have an ordinal measure. But if you were to classify them by the exact number of credits or units they have completed, we would have an interval-ratio variable. Thus you could determine that someone who has completed 64 unit hours has twice as many unit hours as someone who has completed only 32 hours. Therefore, <u>real</u> numbers are associated with each category. Examples of interval-ratio variables are: age (in years), income (in dollars), years married, number of children, years of education.
- It is important to note that levels of measurement are based on a hierarchy. Nominal variables yield the least amount of information; they simply place each person into a category. Ordinal variables yield more information because each person or case is placed into a category and each category can be ranked from high to low. Interval-ratio variables yield the <u>most</u> information; not only are these categories ranked, but the exact distance between the ranks is known and can be measured, thus permitting all mathematical operations to be performed.

4

## MULTIPLE-CHOICE QUESTIONS

1. All boys in Franklin Middle School is an example of a _____.
   a. sample
   b. population
   c. variable
   d. level of measurement

   *Learning Objective: 2*

2. Two types of statistics are_____.
   a. informational and deceptive
   b. descriptive and informational
   c. inferential and deductive
   d. descriptive and inferential

   *Learning Objective: 2*

3. You are interested in average SAT scores of the entering class of freshmen at your school. _____ would be the population; _____ would be the sample.
   a. A selected sub-set of this group; all entering freshmen
   b. All students in the university; a selected sub-set of this group
   c. All entering freshmen at your school; a selected sub-set of this group
   d. All students entering all colleges/universities; all entering freshmen at your school

   *Learning Objective: 2*

4. A researcher wants to study the number of job interviews of graduating seniors in the United States. What is the population?
   a. All recently employed people in the US.
   b. All graduating seniors in the US.
   c. The carefully selected subset of graduating seniors in the US.
   d. All business majors in the US.

   *Learning Objective: 2*

5.  From question 4 above, what would be the key variable for the researcher?
    a. number of graduating seniors
    b. average GPA for the student body
    c. number of job interviews each respondent had
    d. how long it took the average senior to find a job

    *Learning Objective: 3*


6.  The variable "military rank" is a(n) _____ level variable.
    a. nominal
    b. ordinal
    c. interval-ratio
    d. none of the above

    *Learning Objective: 3*


7.  The variable "Do you have a college degree?" is a(n) _____ level variable.
    a. nominal
    b. ordinal
    c. interval-ratio
    d. none of the above

    *Learning Objective: 3*


8.  If you were to rank the levels of measurement from lowest to highest by the amount of information each one yields, what order would you put them in?
    a. ordinal, interval-ratio, nominal
    b. interval-ratio, ordinal, nominal
    c. nominal, ordinal, interval-ratio
    d. nominal, interval-ratio, ordinal

    *Learning Objective: 3*


9.  An adequate scale for measuring a nominal-level variable should be _____.
    a. mutually exclusive, exhaustive, and heterogeneous
    b. mutually exclusive, exhaustive, and homogeneous
    c. mutually inclusive, exhaustive, and homogeneous
    d. none of the above

    *Learning Objective: 3*

10. The variable "Age of children" is a(n) _____ level variable.
   a. nominal
   b. ordinal
   c. interval-ratio
   d. none of the above

   *Learning Objective: 3*

11. Which of the following scales would be best for the variable "ethnicity"?
   a. White, African American, Black
   b. White, Black, Hispanic
   c. Black, non-Black
   d. White, Black, Hispanic, Asian, other

   *Learning Objective: 3*

12. Descriptive statistics are used to _____.
   a. describe a population
   b. describe the relationship between statistics and social science
   c. describe the distribution of a variable
   d. describe the textbook

   *Learning Objective: 2*

13. Inferential statistics are used to _____.
   a. infer the population results to the sample
   b. infer the sample results to the population
   c. infer our knowledge of statistics to social science
   d. infer our knowledge to the instructor

   *Learning Objective: 2*

14. All of the following can be considered variables, except _____.
   a. occupation
   b. marital status
   c. educational attainment
   d. divorced

   *Learning Objective: 3*

15. Independent variables are considered the _____, and dependent variables are considered the _____.
    a. cause, effect
    b. explanation, cause
    c. effect, cause
    d. nominal, explanation

    *Learning Objective: 2*


16. Research is a process where a particular social scientist gathers information in order to _____.
    a. answer questions
    b. examine ideas
    c. test theories
    d. all of above

    *Learning Objective: 1*


17. The numerical information collected and studied by social scientists is called _____.
    a. research
    b. data
    c. theory
    d. statistics

    *Learning Objective: 1*


18. A(An) _____ is a general explanation of the relationships between social phenomena.
    a. hypothesis
    b. research
    c. theory
    d. generalization

    *Learning Objective: 2*

19.  A(An) _____ is a specific statement of the relationships between social
     variables.
     a.  hypothesis
     b.  research
     c.  theory
     d.  generalization

     *Learning Objective: 2*


20.  _____ allow researchers to describe the strength and direction of the
     relationships between two or more variables.
     a.  descriptive statistics
     b.  inferential statistics
     c.  data reduction
     d.  measures of association

     *Learning Objective: 2*



## WORK PROBLEMS

Below are a few of the variables from the General Social Survey (Appendix G in your textbook).
Determine the level of measurement for each variable.

1.  It is alright for a couple to live together without intending to get married.
    1.  Strongly agree
    2.  Agree
    3.  Neither agree nor disagree
    4.  Disagree
    5.  Strongly disagree
    8.  Can't choose

               **Level of Measurement** _____

          *Learning Objective: 3*

2. How often do you attend religious services?
   0. Never
   1. Less than once per year
   2. Once or twice a year
   3. Several times per year
   4. About once a month
   5. 2-3 times a month
   6. Nearly every week
   7. Every week
   8. Several times per week

   **Level of Measurement** _____

   *Learning Objective: 3*

3. How many children have you ever had? Please count all that were born alive at any time (including any from a previous marriage).
   0 - 7. Actual number
   8. Eight or more

   **Level of Measurement** _____

   *Learning Objective: 3*

4. Respondent's highest degree
   0. Less than HS
   1. High school
   2. Assoc/Jr college
   3. Bachelor's
   4. Graduate

   **Level of Measurement** _____

   *Learning Objective: 3*

5. Highest year of school completed
   0-20 Actual number of years

   **Level of Measurement** _____

   *Learning Objective: 3*

6. Respondent's gender
   1. Male
   2. Female

   **Level of Measurement** _____

   *Learning Objective: 3*


7. Is there any area right around here – that is, within a mile – where you would be afraid to walk alone at night?
   1. Yes
   2. No

   **Level of Measurement** _____

   *Learning Objective: 3*


8. How often do you read the newspaper?
   1. Everyday
   2. A few times a week
   3. Once a week
   4. Less than once a week
   5. Never

   **Level of Measurement** _____

   *Learning Objective: 3*


9. In 2000, you remember that Gore ran for president on the Democratic ticket against Bush for the Republicans and Nader as an Independent. Did you vote for Gore, Bush, or Nader. (Includes only those who said they voted in the 2000 election.)
   1. Gore
   2. Bush
   3. Nader
   4. Other
   6. No presidential vote

   **Level of Measurement** _____

   *Learning Objective: 3*

10. Are you currently married, widowed, divorced, separated, or have you never been married?
    1. Married
    2. Widowed
    3. Divorced
    4. Separated
    5. Never married

    **Level of Measurement** _____

    *Learning Objective: 3*

For the next questions, identify the independent and the dependent variables.

11. A researcher interested in studying the relationship between religiosity and prejudice finds that higher levels of religiosity tend to produce higher levels of racial prejudice.

    **Independent Variable** _____

    **Dependent Variable** _____

    *Learning Objective: 2*

12. Another researcher seeks to test the relationship between education and voting; more specifically, the researcher seeks to determine if higher educational levels lead to increases in voting behavior.

    **Independent Variable** _____

    **Dependent Variable** _____

    *Learning Objective: 2*

13. A third researcher studies whether cities with more immigrants have higher or lower levels of unemployment.

    **Independent Variable** _____

    **Dependent Variable** _____

    *Learning Objective: 2*

## ANSWER KEY

### MULTIPLE-CHOICE QUESTIONS

| | | | | | | |
|---|---|---|---|---|---|---|
| 1. | b | (p. 16) | 11. | d | (p. 19) |
| 2. | d | (p. 15-16) | 12. | c | (p. 15) |
| 3. | c | (p. 16) | 13. | b | (p. 16-17) |
| 4. | b | (p. 16) | 14. | d | (p. 11) |
| 5. | c | (p. 11) | 15. | a | (p. 11) |
| 6. | b | (p. 17) | 16. | d | (p. 9) |
| 7. | a | (p. 17) | 17. | b | (p. 9) |
| 8. | c | (p. 17) | 18. | c | (p. 10) |
| 9. | b | (p. 19) | 19. | a | (p. 12) |
| 10. | c | (p. 21) | 20. | d | (p. 15-16) |

### WORK PROBLEMS

1.  This variable is a scale item, sometimes referred to as a Likert item scale, where every response is ranked on a 1 - 5 scale ranging from strongly agree to strongly disagree. Scale item are ordinal variables because every response can be rank-ordered on the degree to which the respondent agrees or disagrees with the statement.

2.  This variable is ordinal because it can be ordered by the amount of time the respondent attends religious services per year. We know that someone who attends every week, attends more often than someone who attends only once or twice per year, which is more often than someone who attends less than once per year.

3.  Because the question asks for the actual number, this variable is interval-ratio. We know that having two children is two times more than having one child; the distance between the numbers is known. As such, this variable yields the most information and more complex mathematical operations can be performed on this type of variable.

4.  This variable is ordinal, because we can rank-order the categories. We know that having a college degree is higher than having a high school diploma, but we cannot say for certain (with the amount of detail given in these categories) that having a bachelor's degree is four times higher than having a high school diploma. If we want to get this type of information, we can transform these categories into their corresponding years of education, such as a high school diploma entails 12 years of education, associate degree entails 14 years of education, etc. Unfortunately, in doing so, you will have to assign values for those unknowns such as "less than high school" and "graduate degree." Someone with less than high school might have five years of education, while others might have eleven. Be careful of these problems whenever you attempt to change the level of measurement; you might lose more data than it's worth!

5. This variable asks the respondent for the actual number of years of education rather than the degrees he/she might have acquired; therefore it is measured at the interval-ratio level. Compare this question with Question 4. Can you see that we obtain more information when we ask the interval-ratio version of this question rather than the ordinal level? Here, we don't have to make "guesstimates" about the number of years, but instead we know the exact value. Therefore, if someone dropped out of high school during his/her senior year, she/he would have checked "Less than HS" in question 4, putting him/her in the same category with others who perhaps did not continue after middle school. By asking for number of years, though, the respondent could mark 11.5 years, thus distinguishing him/herself from someone with 8 years of education.

6. Gender is a nominal-level variable: being either male or female is simply a categorization. Gender categories cannot be ranked.

7. This is a nominal-level variable. All yes/no, favor/oppose, agree/disagree (dichotomous, or two category, variables) are nominal level. Perhaps a better way to ask this question would be to make it an ordinal level one by asking to what degree the respondent is fearful: not at all, a little bit, quite a bit, extremely. Instead, the researchers decided to keep it simple by asking it as a yes/no question, thus making it a nominal-level variable.

8. This variable ranks the amount of time people read the newspaper, thus making it an ordinal-level variable. Someone who reads the paper every day, reads it more often than someone who reads it a few times per week. This is not an interval-ratio level variable, because we don't have the exact distance between the categories.

9. This variable is nominal because it simply places each respondent in a category based on who he/she voted for. It does not make assumptions regarding strength of his/her political beliefs; it simply classifies people into one category or another.

10. This variable is nominal because these categories cannot be ordered or ranked. Being married is not higher than not being married. Therefore, each respondent can be placed in one and only one unrankable category.

11. Because the researcher argues that changes in religiosity will produce changes in prejudice, the independent variable is religiosity, and the dependent variable is prejudice.

12. Because educational attainment tends to precede voting behavior, education is the independent variable, and voting behavior is the dependent variable.

13. The independent variable is the number of immigrants in a city, and the dependent variable is the city's unemployment rate.

# CHAPTER TWO
## Basic Descriptive Statistics:
### Percentages, Ratios and Rates, Frequency Distributions

*Learning Objectives: By the end of this chapter, you will be able to:*

1. *Explain the purpose of descriptive statistics in making data comprehensible.*
2. *Compute and interpret percentages, proportions, ratios, rates, and percentage change.*
3. *Construct and analyze frequency distributions for variables at each of the three levels of measurement.*

## KEY TERMS

- ✓ Cumulative frequency (p. 43)
- ✓ Cumulative percentage (p. 43)
- ✓ Frequency distribution (p. 37)
- ✓ Midpoint (p. 42)
- ✓ Percentage (p. 30)
- ✓ Percentage change (p. 35)
- ✓ Proportion (p. 30)
- ✓ Rate (p. 34)
- ✓ Ratio (p. 33)

## CHAPTER SUMMARY

You have now been armed with the basic knowledge necessary to begin data analysis. Keep in mind that data analysis belongs to one of two groups: descriptive (describing the distribution of a variable) or inferential (inferring your sample results to the population). You always want to begin your data analysis by describing the distribution of a variable, such as what percentage of your sample is male. For this we use basic descriptive statistics, such as percentages, ratios, rates, and frequency tables.

These basic descriptive techniques display our data in easily readable format, enabling us to look at the distribution of each variable. The first step in this process is to make a frequency distribution table. We can do this with all three types of variables (nominal, ordinal, interval-ratio), although the process is a bit different for interval-ratio variables. A next step in the process of describing a variable is to calculate any of the various descriptive statistics described in this chapter, including proportions, percentages, ratios, rates, and/or the percent of change over time.

# CHAPTER OUTLINE

## 2.1  Percentages and Proportions.

- Percentages and proportions standardize raw data, and because of this, they provide a useful frame of reference for research results.  They are particularly useful when comparing unequally sized groups.
- Basically it works like this:  let's say you have 100 respondents, 60 of whom are female and 40 of whom are male.  We want to display this distribution across these two categories, such as in Table 2.1 (below).  Looking at this table we can easily see that the majority of the respondents are female (60 out of 100, or 60%).  This table, like all frequency distributions, gives the frequency as well as the percentage and/or the proportion in each category.

TABLE 2.1   FREQUENCY DISTRIBUTION OF GENDER

| Gender | Frequency (f) | Percentage (%) | Proportion (p) |
|---|---|---|---|
| (1)  Male | 40 | 40% | 0.40 |
| (2)  Female | 60 | 60% | 0.60 |
| N= | 100 | 100% | 1.00 |

To find the percentage for each category, we use:

$$Percentage, \; \% = \left( \frac{f}{N} \right) \times 100$$

where   f  = frequency, or the number of cases in a category
N = number of cases in all categories

From the example above, the percentage of females is:

$$Percentage, \; \% = \left( \frac{60}{100} \right) \times 100 = 60\%$$

To find the proportion, we use:

$$Proportion, \; p = \left( \frac{f}{N} \right)$$

where   f  = frequency, or the number of cases in a category
N = number of cases in all categories

From the example above, the proportion of females is:

$$Proportion,\ p = \left(\frac{60}{100}\right) = 0.60$$

- The only difference between percentages and proportions is their base number: percentages are based on 100, while proportions are based on 1. They yield the same information.
- Additional guidelines for the use of percentages and proportions include:
  - Report frequencies, instead of proportions or percentages, when describing a small number of cases.
  - Report the number of cases whenever a proportion or percentage is reported.
  - Proportions and percentages can be calculated for nominal, ordinal and interval-ratio variables.

## 2.2   Ratios, Rates, and Percentage Change.

- Proportions, rates, and ratios can be confusing. If you're not sure which one to use, maybe this will help: if you are interested in the number of females attending your college compared to males (that is, when you want to compare the relative size of any two categories of a variable), then a ratio is most appropriate. Let's say there are 1200 female and 900 male students at your school. The ratio of females to males would be:

$$Ratio = \frac{f_1}{f_2} = \frac{1200}{900} = 1.33$$

Therefore, there are 1.33 females for every male at your college.

- But what if you're interested in the graduation rate for your college? (A rate is appropriate when you're interested in the occurrence of a particular event, such as graduations, births, deaths, pregnancies, homicides, etc.) If you know that 320 of the initial 400 freshmen graduated within four years, then the graduation *rate* would be:

$$Rate = \frac{Actual\ Occurrences}{Possible\ Occurrences} \times (number)$$

$$Rate = \frac{Number\ Graduated}{Number\ of\ Freshmen} \times 1000 = \frac{320}{400} \times 1000 = 800$$

Therefore, for every 1000 incoming freshmen, you can expect 800 to graduate within four years.

The "number" used in the rate formula is used to make sure that the final result is a whole number. This number is often one thousand (1,000), ten thousand (10,000), or one hundred thousand (100,000), depending on the relative rarity of the event being studied.

Of course both ratios and rates can be viewed in terms of proportions or percentages, too. For example, there are 0.43 (43%) males in the college [p = 900/2100; % = 100×(900/2100)], and 0.80 (80%) graduate within four years [p=320/400; % = 100×(320/400)].

- Percentage change can be a very useful statistic especially when thinking about such things as the change in wins for your college football team or, more generally, when you want to determine how much something has changed (increased or decreased) over time. Let's say that last year the team won 15 games, whereas this year they won only 12 games. Using the percentage change equation we see that the team's win record has decreased by 25%.

$$Percentage\ Change = \left(\frac{f_2 - f_1}{f_1}\right) \times 100 = \left(\frac{12 - 15}{12}\right) \times 100 = -25\%$$

## 2.3 Frequency Distributions: Introduction.
- A frequency distribution table is a table that displays the frequency (or number) of respondents in each category for that variable. Such tables are very useful ways of presenting data. In fact, the construction of a frequency table is usually the first step in the analysis of any variable.
- As a general rule, the categories of a variable represented in a frequency distribution table should be mutually exhaustive, exclusive, and homogenous.

## 2.4 Frequency Distributions for Variables Measured at the Nominal and Ordinal Levels.
- The construction of a frequency distribution table for a nominal variable is fairly straightforward. Each category of the variable is listed in the left-most column of the table. Then, the number of occurrences for each category are listed in the next column, labeled "Frequency." Next, the total number of cases is reported in the bottom row of the table. Finally, the researcher may also add a percentage and/or proportion column to the table.
  - Usually, each category of the variable is listed in the frequency table, but occasionally the researcher will want to combine or collapse categories depending on the particular research question.
- Frequency distribution tables for ordinal variables are constructed in the same way. However, the categories of the variable should be presented in the order of their ranking.

## 2.5 Frequency Distributions for Variables Measured at the Interval-Ratio Level.

- When working with interval-ratio variables, the frequency distribution tables are created differently. Think of all the different ages of the other students in your statistics class. If you have a class of 50 students, you probably have some students who are 18 years old, perhaps one or two who are under 18, a few who are 19, several who are 20, some who are between 21 and 24, perhaps some who are around 30, and some who are older than 30. To put all of these scores into a frequency distribution table, you might have quite a lot of categories. It's possible, though, to break these categories down into more manageable sizes, or class intervals such as 17-19 years of age, 20-22, 23-25, 26-28, etc. This would enable you to see the distribution more clearly. The size of the class intervals (or even the use of them at all) is up to you depending on how much information you want to convey.

  - When used, class intervals should be of equal size for ease in interpretation. (Although there are some exceptions to this "rule," as in the case of reporting incomes. See the example in your text book.) Under some circumstances (for example, when there are a few cases with either very high or very low values compared to the other cases), the researcher will want to use open-ended intervals, such as "less than 20" or "30 and older."

  - To determine the size or width of class intervals, first determine how many intervals you want to use, or $k$. (Ten intervals is the convention, but sometimes the researcher will use more or fewer intervals depending on the variable.) Find the range of the values, $R$, on the variable by subtracting the lowest value from the highest value. Finally, calculate the interval width, $i$, by dividing $R$ by $k$ and then rounding to a convenient whole number (in symbols, $i = R/k$). The lowest interval usually starts at or below the lowest value in the variable's distribution.

- Midpoints are the points exactly in the middle of the intervals; they are found by dividing the sum of the upper and lower class limits by two. If the class limits are 17 - 19, the sum is 36; divided by two, the midpoint is 18. (See Table 2.2 below.)

### TABLE 2.2   FREQUENCY DISTRIBUTION OF AGE IN CLASS

| Ages | Frequency | Cumulative Frequency | Midpoint | Percent | Cumulative Percent |
|------|-----------|----------------------|----------|---------|--------------------|
| 17-19 | 8 | 8 | 18 | 16 | 16 |
| 20-22 | 21 | 29 | 21 | 42 | 58 |
| 23-25 | 11 | 40 | 24 | 22 | 80 |
| 26-28 | 8 | 48 | 25 | 16 | 96 |
| 29-31 | 2 | 50 | 30 | 4 | 100 |
| | N = 50 | | | 100% | |

- Cumulative frequencies and cumulative percentages are also used in frequency distributions for interval-ratio variables. Cumulative frequencies are the sum of the frequencies down the columns. Thus if there are 8 students between the ages of 17 and 19, the first cumulative frequency would be 8. If there are 21 students between the ages of 20 and 22, then the cumulative frequency would be 8 +21, or 29. Cumulative percentages work the same way, but you sum the percentage column instead of the frequency column.
  - Both of these measures make it easier to view the distribution. With a quick look at the cumulative percentage column, you can tell where 80% of the cases fall below (in the above example, 23-25 and below); with the cumulative frequency/percentage column you can also quickly find the median (described in chapter 4) for the distribution.

**2.6 Constructing Frequency Distributions for Interval-Ratio Level Variables: A Review.**

**MULTIPLE-CHOICE QUESTIONS**

1.    A frequency distribution table is _____.
       a. a table with the total number of respondents
       b. a table with the number of respondents by category
       c. the percentage or proportion of respondents by category
       d. all of these choices

*Learning Objective: 3*

2.    For nominal variables, the frequency distribution should have _____.
       a. cumulative frequencies
       b. cumulative percentages or proportions
       c. percentages or proportions
       d. all of these choices

*Learning Objective: 3*

3.    For ordinal variables, the frequency distribution should have _____.
       a. cumulative frequencies
       b. cumulative percentages or proportions
       c. percentages or proportions
       d. all of these choices

*Learning Objective: 3*

4.  For interval-ratio variables, the frequency distribution should have _____.
    a. cumulative frequencies
    b. cumulative percentages or proportions
    c. percentages or proportions
    d. all of these choices

    *Learning Objective: 3*

5.  Percentages are based on _____.
    a. one
    b. five
    c. fifty
    d. one hundred

    *Learning Objective: 2*

6.  Proportions are based on _____.
    a. one
    b. five
    c. fifty
    d. one hundred

    *Learning Objective: 2*

7.  Ratios are _____.
    a. the number of actual occurrences of some phenomenon or trait divided by the number of possible occurrences per some unit of time
    b. the number of cases in one category divided by the number of cases in another category
    c. the number of cases in one category of a variable divided by the number of cases in all categories of the variable
    d. the horizontal dimension of a table

    *Learning Objective: 2*

8.  Rates are _____.
    a.  the number of actual occurrences of a phenomenon or trait divided by the number of possible occurrences per a unit of time
    b.  the number of cases in one category divided by the number of cases in another category
    c.  the number of cases in one category of a variable divided by the number of cases in all categories of the variable
    d.  the horizontal dimension of a table

    *Learning Objective: 2*

9.  _____ are usually preferred over _____ because they are easier to read and comprehend and because they are better for the comparison of groups.
    a.  frequencies, percentages
    b.  proportions, percentages
    c.  percentages, frequencies
    d.  rates, ratios

    *Learning Objective: 2*

10. When a distribution has a small number of cases, it is usually preferable to report

    _____.
    a.  percentages
    b.  frequencies
    c.  cumulative proportions
    d.  percentage change

    *Learning Objective: 2*

11. Whenever you report a proportion and/or percentage, you should always report the

    _____ too.
    a.  number of categories
    b.  level of measurement
    c.  number of observations
    d.  percentage change

    *Learning Objective: 2*

12. Frequency distribution tables can be constructed for _____ level variables.
    a. ordinal
    b. nominal
    c. interval-ratio
    d. all of these choices

    *Learning Objective: 3*

13. Percentages and proportions can be calculated for _____ level variables.
    a. ordinal
    b. nominal
    c. interval-ratio
    d. all of these choices

    *Learning Objective: 2*

14. _____ are useful and economical ways of expressing the relative size of two groups.
    a. ratios
    b. rates
    c. proportions
    d. percentages

    *Learning Objective: 2*

15. _____ is a very useful statistic because it tells us whether a variable has increased or decreased over time.
    a. percentage
    b. percentage change
    c. rate
    d. ratio

    *Learning Objective: 2*

16. The construction of a _____ is the first step in any analysis.
    a. cumulative frequency
    b. midpoint
    c. frequency distribution
    d. class interval

*Learning Objective: 3*


17. When determining class intervals for a distribution, use this procedure.
    a. divide the range of values by the number of intervals and round to a whole number
    b. determine the range of values and round to a whole number
    c. divide the number of intervals by the range of values and round to a whole number
    d. multiply the range of values by the number of intervals

*Learning Objective: 3*


18. By convention, we usually use _____ class intervals in a frequency distribution
    a. 1
    b. 5
    c. 10
    d. 12

*Learning Objective: 3*


19. When calculating a midpoint for a class interval, we _____.
    a. subtract the lower limit from the upper limit and divide by two
    b. multiply the upper and lower limits and divide by two
    c. subtract the lower limit from the upper limit
    d. add the upper and lower limits and divide by two

*Learning Objective: 3*


20. The cumulative percentage for a category tells us _____.
    a. the percentage of cases in a particular category
    b. the percentage of cases in and below a particular category
    c. the percentage of cases above a particular category
    d. the relative change in a category over time

*Learning Objective: 3*

# WORK PROBLEMS

1.  A researcher collected data on attitudes toward abortion. The first step in analyzing the data was to make the frequency distribution table below.

| Attitude Toward Abortion | Frequency (f) |
|---|---|
| Strongly favor | 42 |
| Favor | 57 |
| Oppose | 52 |
| Strongly oppose | 43 |
|  | N = 194 |

a.  Complete the table by adding a percentage column.

b.  What conclusions can you make regarding the data?

*Learning Objective: 3*

2.  The data below represent the age at which a sample of newlyweds got married.

| 17 | 25 | 27 | 22 | 35 | 23 | 27 | 28 | 19 | 19 | 22 |
|---|---|---|---|---|---|---|---|---|---|---|
| 18 | 18 | 24 | 31 | 16 | 18 | 26 | 21 | 21 | 23 | 36 |
| 24 | 19 | 22 | 27 | 21 | 23 | 25 | 22 | 22 | 20 | |

a.  Make a frequency distribution table from this data. (Use an interval width of 2 years.) Include frequency, cumulative frequency, percentage, and cumulative percentage columns.

b.  What conclusions can you make?

*Learning Objective: 3*

3.  The data for age first married was broken down by gender.

| Males | | Females | |
|---|---|---|---|
| 24 | 27 | 17 | 18 |
| 36 | 23 | 18 | 23 |
| 21 | 19 | 22 | 19 |
| 25 | 26 | 21 | 27 |
| 19 | 25 | 35 | 22 |
| 23 | 22 | 16 | 21 |
| 27 | 28 | 18 | 22 |
| 24 | 31 | 19 | 20 |

25

a. Make a frequency distribution table for each group (male and female) from this data. (Use an interval width of 2 years.) Include frequency, cumulative frequency, percentage, and cumulative percentage columns.

b. What conclusions can you make?

*Learning Objective: 3*

4. The following data report the overall satisfaction with the police response in one inner-city neighborhood.

| Satisfaction with Police | Frequency (f) |
| --- | --- |
| Very satisfied | 25 |
| Satisfied | 32 |
| Dissatisfied | 38 |
| Very dissatisfied | 27 |
| | N = 122 |

a. Complete the frequency distribution table by adding a percentage column.

b. What conclusions can you make regarding satisfaction with police?

*Learning Objective: 3*

5. The following questions were asked of 20 respondents.

Sex:  1 Male
      2 Female

Education (given in actual years)

Should marijuana be legalized?
      1 Yes
      2 No

The data for these questions appear at the top of the next page.

| Case | Sex | Education | Legalize Marijuana |
|:---:|:---:|:---:|:---:|
| 1 | 1 | 12 | 1 |
| 2 | 2 | 12 | 2 |
| 3 | 1 | 14 | 2 |
| 4 | 1 | 15 | 1 |
| 5 | 2 | 11 | 1 |
| 6 | 2 | 9 | 2 |
| 7 | 1 | 16 | 1 |
| 8 | 1 | 16 | 1 |
| 9 | 2 | 14 | 2 |
| 10 | 2 | 12 | 1 |
| 11 | 2 | 11 | 1 |
| 12 | 1 | 12 | 1 |
| 13 | 1 | 12 | 2 |
| 14 | 1 | 17 | 1 |
| 15 | 2 | 18 | 2 |
| 16 | 1 | 10 | 2 |
| 17 | 1 | 14 | 1 |
| 18 | 2 | 15 | 2 |
| 19 | 1 | 9 | 2 |
| 20 | 1 | 11 | 1 |

Construct a frequency distribution for each variable. (Use an interval width of 2 years for the education variable.) For each, add the appropriate columns depending on each variable's level of measurement.

*Learning Objective: 3*

6.    Using the data from problem 5, interpret the cumulative percentages for education.

*Learning Objective: 2*

7.    The following data are from a quality of life scale administered to hospitalized respondents. Low scores indicate low quality of life, high scores indicate high quality of life. Assuming the data to be interval-ratio, construct a frequency distribution and write a conclusion. (Use an interval width of 2 points.)

**Quality of Life**

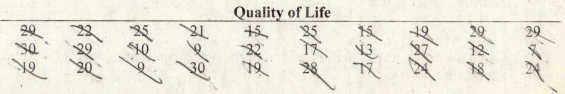

| 29 | 22 | 25 | 21 | 15 | 25 | 15 | 19 | 29 | 29 |
| 30 | 29 | 10 | 9 | 22 | 17 | 13 | 27 | 12 | 7 |
| 19 | 20 | 9 | 30 | 19 | 28 | 17 | 24 | 18 | 24 |

*Learning Objective: 3*

27

8.    The city council is trying to get an idea of the religious background of the city's residents.

   The following data were collected. What would you conclude about the religious background of the community?

| Religious Affiliation | Frequency (f) |
|---|---|
| Protestant | 240 |
| Catholic | 467 |
| Jewish | 139 |
| Other | 86 |
| None | 22 |
| | N = 954 |

*Learning Objective: 3*

9.    Using the data from problem 8, calculate the ratio of Catholic residents to Protestant residents.

*Learning Objective: 2*

10.   Using the data from problem 8, calculate the proportion of city residents who are not Protestant, Catholic, or Jewish.

*Learning Objective: 2*

11.   Complete the frequency distribution table by adding the appropriate column(s) for the variable's level of measurement.

| Highest Educational Degree | Frequency (f) |
|---|---|
| Less than high school | 8 |
| High school | 27 |
| Associates or junior college | 15 |
| Bachelors | 12 |
| Graduate | 3 |
| | N = 65 |

*Learning Objective: 3*

12.   Using the data from problem 11, calculate the percentage that have at least some college education.

*Learning Objective: 2*

28

13. In a city of 750,000 people, the frequency of unwed pregnancies in a one-year period was 1,875. What is the crude unwed pregnancy rate for this city?

*Learning Objective: 2*

14. The following data about drinking patterns were collected at a large university.

| Frequency of Drinking | Social Sciences | Major Arts and Humanities | Physical Sciences |
|---|---|---|---|
| Less than once/week | 12 | 8 | 6 |
| Once or twice/week | 17 | 5 | 8 |
| Three to five times/week | 13 | 4 | 6 |
| Five to seven times/week | 6 | 2 | 3 |
| More than seven times/week | 2 | 0 | 2 |
| | N = 50 | N = 19 | N = 25 |

a. Students in which major drink proportionally more than the other majors?

b. What is the ratio of heavy drinkers (seven or more times per week) to light drinkers (less than once per week) by majors?

*Learning Objective: 2*

15. In a city with a population of one million, there were 516 homicides in the past year. What is the crude homicide rate?

*Learning Objective: 2*

## SPSS or MicroCase WORK PROBLEMS

The following work problems can be completed using either SPSS or MicroCase. The solutions will be the same regardless of which one you use. Your instructor may have a preference, so be sure to check with him/her prior to beginning these problems.

1. Create frequency distributions for the following variables: *abany*, *happy*, and *degree*.

2. From the data output above, describe abortion attitudes.

3. From the data output above, describe general happiness levels.

4. From the data output above, describe educational attainments.

5.  Recode the variable *age* into a new variable called *age4* so that people who are 18 to 34 are in one category (use the label "18-34"), people who are 35 to 49 are in a second category (use the label "35-49"), people who are 50 to 64 are in a third category (use the label "50-64"), and people who are 65 or older are in a fourth category (use the label "65+"). (Hint: The oldest age in the sample is 89.) Label all categories of the *age4* variable, and output frequency distributions of both *age* and *age4*. Check that the numbers correspond correctly.

6.  Recode the variable *degree* into a new variable called *degree2* so that people who have at least a four-year college degree are in one category (use the label "college grad") and everyone else is in a second category (use the label "less than college"). Label all categories of the *degree2* variable, and output frequency distributions of both variables Check that the numbers correspond correctly. Compare the two distributions.

## ANSWER KEY

### MULTIPLE-CHOICE QUESTIONS

| | | | | | |
|---|---|---|---|---|---|
| 1. | d | (p. 37) | 11. | c | (p. 32) |
| 2. | c | (p. 39) | 12. | d | (p. 37-38) |
| 3. | c | (p. 40) | 13. | d | (p. 32) |
| 4. | d | (p. 40-42) | 14. | a | (p. 33) |
| 5. | d | (p. 30) | 15. | b | (p. 35) |
| 6. | a | (p. 30) | 16. | c | (p. 37) |
| 7. | b | (p. 33) | 17. | a | (p. 41-42) |
| 8. | a | (p. 34) | 18. | c | (p. 46) |
| 9. | c | (p. 30-31) | 19. | d | (p. 42-43) |
| 10. | b | (p. 32) | 20. | b | (p. 43-44) |

### WORK PROBLEMS

1a.

| Attitude Toward Abortion | Frequency (f) | Percentage (%) |
|---|---|---|
| Strongly favor | 42 | 22 |
| Favor | 57 | 29 |
| Oppose | 52 | 27 |
| Strongly oppose | 43 | 22 |
| | N = 194 | 100 % |

1b.  The majority of respondents do not strongly favor or strongly oppose abortion, but tend to fall fairly evenly between the two middle categories.

2a.

| Age First Married | Frequency (f) | Cumulative Frequency | Percentage (%) | Cumulative Percentage |
|---|---|---|---|---|
| 16 – 17 | 2 | 2 | 6.3 | 6.3 |
| 18 – 19 | 6 | 8 | 18.8 | 25.1 |
| 20 – 21 | 4 | 12 | 12.5 | 37.6 |
| 22 – 23 | 8 | 20 | 25.0 | 62.6 |
| 24 – 25 | 4 | 24 | 12.5 | 75.1 |
| 26 – 27 | 4 | 28 | 12.5 | 87.6 |
| 28 – 29 | 1 | 29 | 3.1 | 90.7 |
| 30 and older | 3 | 32 | 9.4 | 100.1 |
|  | N = 32 |  | 100.1 % |  |

2b. There are more people who got married between the ages of 22 and 23 than any other category (25%), and the majority of the sample (62.6%) was married by the age of 23.

3a. **MEN**

| Age First Married | Frequency (f) | Cumulative Frequency | Percentage (%) | Cumulative Percentage |
|---|---|---|---|---|
| 16 – 17 | 0 | 0 | 0.0 | 0.0 |
| 18 – 19 | 2 | 2 | 12.5 | 12.5 |
| 20 – 21 | 1 | 3 | 6.3 | 18.8 |
| 22 – 23 | 3 | 6 | 18.8 | 37.6 |
| 24 – 25 | 4 | 10 | 25.0 | 62.6 |
| 26 – 27 | 3 | 13 | 18.8 | 81.4 |
| 28 – 29 | 1 | 14 | 6.3 | 87.7 |
| 30 and older | 2 | 16 | 12.5 | 100.2 |
|  | N = 16 |  | 100.2 % |  |

**WOMEN**

| Age First Married | Frequency (f) | Cumulative Frequency | Percentage (%) | Cumulative Percentage |
|---|---|---|---|---|
| 16 – 17 | 2 | 2 | 12.5 | 12.5 |
| 18 – 19 | 5 | 7 | 31.3 | 43.8 |
| 20 – 21 | 3 | 10 | 18.8 | 62.6 |
| 22 – 23 | 4 | 14 | 25.0 | 87.6 |
| 24 – 25 | 0 | 14 | 0.0 | 87.6 |
| 26 – 27 | 1 | 15 | 6.3 | 93.9 |
| 28 – 29 | 0 | 15 | 0.0 | 93.9 |
| 30 and older | 1 | 16 | 6.3 | 100.2 |
|  | N = 16 |  | 100.2 % |  |

3b.    The data in this sample indicate that the majority of males marry at an older age than do the females. Over half of the female respondents (62.6%) were married by the age of 21, compared to only 18.8% of males.

4a.

| Satisfaction with Police | Frequency (f) | Percentage (%) |
|---|---|---|
| Very satisfied | 25 | 21 |
| Satisfied | 32 | 26 |
| Dissatisfied | 38 | 31 |
| Very dissatisfied | 27 | 22 |
| | N = 122 | 100 % |

4b.    There is no consensus on attitudes toward police response, although slightly more respondents are dissatisfied than satisfied.

5.

| Sex | Frequency (f) | Percentage (%) |
|---|---|---|
| Male | 12 | 60 |
| Female | 8 | 40 |
| | N = 20 | 100 % |

| Education | Frequency (f) | Cumulative Frequency | Percentage (%) | Cumulative Percentage |
|---|---|---|---|---|
| 9 – 10 | 3 | 3 | 15 | 15 |
| 11 – 12 | 8 | 11 | 40 | 55 |
| 13 – 14 | 3 | 14 | 15 | 70 |
| 15 – 16 | 4 | 18 | 20 | 90 |
| 17 – 18 | 2 | 20 | 10 | 100 |
| | N = 20 | | 100 % | |

| Legalize Marijuana | Frequency (f) | Percentage (%) |
|---|---|---|
| Yes | 11 | 55 |
| No | 9 | 45 |
| | N = 20 | 100 % |

6. 15% of respondents have 9-10 years of education. 55% of respondents have 11-12 years of education or less. 70% of respondents have 13-14 years of education or less. 90% of respondents have 15-16 years of education or less. 100% of respondents have 17-18 years of education or less.

7.

| Quality of Life | Frequency (f) | Cumulative Frequency | Percentage (%) | Cumulative Percentage |
|---|---|---|---|---|
| 7 – 9 | 4 | 4 | 13.3 | 13.3 |
| 10 – 12 | 2 | 6 | 6.7 | 20.0 |
| 13 – 15 | 3 | 9 | 10.0 | 30.0 |
| 16 – 18 | 3 | 12 | 10.0 | 40.0 |
| 19 – 21 | 5 | 17 | 16.7 | 56.7 |
| 22 – 24 | 4 | 21 | 13.3 | 70.0 |
| 25 – 27 | 3 | 24 | 10.0 | 80.0 |
| 28 – 30 | 6 | 30 | 20.0 | 100.0 |
| | N = 30 | | 100 % | |

The majority of the respondents scored lower than 21 on the Quality of Life scale (56.7%), but 20% of the respondents scored between 28 and 30 on the scale.

8. Almost half of the sample is Catholic (49%), another 25% is Protestant.

9. In order to find the ratio of Catholic residents to Protestant residents, divide the number of Catholics by the number of Protestants.

Ratio = (467 / 240) = 1.95

There are 1.95 Catholic residents for every Protestant resident.

10. In order to find the proportion of residents who are not Protestant, Catholic, or Jewish, add the number of residents who have no or another affiliation and then divide by the total number of residents.

Proportion = (86 + 22 / 954) = 0.11

Residents with no or another religious affiliation represent 0.11 of the city.

11.

| Highest Educational Degree | Frequency (f) | Percentage (%) |
|---|---|---|
| Less than high school | 8 | 12.3 |
| High school | 27 | 41.5 |
| Associates or junior college | 15 | 23.1 |
| Bachelors | 12 | 18.5 |
| Graduate | 3 | 4.6 |
| | N = 65 | 100.0 % |

12. In order to find the percentage of cases with at least some college, add the number of people in the associates, bachelors, and graduate categories, divide by the total number of people, and multiply by 100.

Percentage = $(15 + 12 + 3 / 65) \times 100 = 46.2\%$

46.2% of cases have at least some college education.

13. In order to find the rate of unwed pregnancies divide the total number of unwed pregnancies by the population, and multiply by 1,000:

Crude pregnancy rate = $(1875 / 750,000) \times 1000 = 2.5$

Therefore there are 2.5 unwed pregnancies for every 1,000 people.

14a. In order to find which major drinks more, you need to calculate the proportions for each category by each major. Proportionally, physical science majors drink more than seven times per week (0.8) than do social science (0.04) and arts/humanity majors (0).

14b. For every light social science drinker, there is 0.16 heavy drinkers (ratio = 2/12); there are no heavy drinkers in the arts/humanity majors; and for the physical sciences, for every light drinker there is 0.33 heavy drinkers (ratio = 6/25).

15. Crude homicide rate = $(516 / 1,000,000) \times 10,000 = 5.16$

Therefore there are 5.16 homicides for every 10,000 people.

## SPSS or MicroCase WORK PROBLEMS

If you had problems with any of these questions, be sure to re-read the appropriate sections at the end of chapter 2 in your textbook.

1. **ABANY: ABORTION IF WOMAN WANTS FOR ANY REASON**

| | | Frequency | Percent | Valid Percent | Cumulative Percent |
|---|---|---|---|---|---|
| Valid | YES | 260 | 17.3 | 40.4 | 40.4 |
| | NO | 384 | 25.6 | 59.6 | 100.0 |
| | Total | 644 | 42.9 | 100.0 | |
| Missing | NAP | 836 | 55.7 | | |
| | NA | 20 | 1.3 | | |
| | Total | 856 | 57.1 | | |
| Total | | 1500 | 100.0 | | |

**DEGREE: RS HIGHEST DEGREE**

| | | Frequency | Percent | Valid Percent | Cumulative Percent |
|---|---|---|---|---|---|
| Valid | LT HIGH SCHOOL | 222 | 14.8 | 14.8 | 14.8 |
| | HIGH SCHOOL | 768 | 51.2 | 51.2 | 66.0 |
| | JUNIOR COLLEGE | 114 | 7.6 | 7.6 | 73.6 |
| | BACHELOR | 262 | 17.5 | 17.5 | 91.1 |
| | GRADUATE | 133 | 8.9 | 8.9 | 100.0 |
| | Total | 1499 | 99.9 | 100.0 | |
| Missing | NA | 1 | .1 | | |
| Total | | 1500 | 100.0 | | |

**HAPPY: GENERAL HAPPINESS**

| | | Frequency | Percent | Valid Percent | Cumulative Percent |
|---|---|---|---|---|---|
| Valid | VERY HAPPY | 329 | 21.9 | 32.9 | 32.9 |
| | PRETTY HAPPY | 526 | 35.1 | 52.7 | 85.6 |
| | NOT TOO HAPPY | 144 | 9.6 | 14.4 | 100.0 |
| | Total | 999 | 66.6 | 100.0 | |
| Missing | NAP | 497 | 33.1 | | |
| | NA | 4 | .3 | | |
| | Total | 501 | 33.4 | | |
| Total | | 1500 | 100.0 | | |

2. To summarize each of the distributions, you want to read down the column marked "Valid Percent." For the variable "Do you favor abortion if a woman wants it for any reason?" (*abany*), 40.4% stated yes and 59.6% stated no. Therefore, a majority of the sample disagreed that a woman should be able to have an abortion for any reason.

3. For the variable "general happiness" (*happy*), 32.9% claim to be very happy, 52.7% are pretty happy, and 14.4% aren't too happy. Therefore, almost everyone (32.9% + 52.7% = 88.6%) reports being happy.

4. For the variable "respondent's highest degree" (*degree*), 14.8% have less than a high school education, 51.2% have a high school diploma, 7.6% have a junior college education, 17.5% have a bachelor's degree, and 8.9% have a graduate degree. Therefore, a majority of the sample are high school graduates.

5. The following frequency distributions are for *age4* (age collapsed into four categories) and *age*. Your numbers should add the same as mine. If not, check that the missing values in original variable are still missing in the new variable. (The *age* variable has a lot of categories, so it extends across several pages.)

**AGE: AGE OF RESPONDENT**

|        |    | Frequency | Percent | Valid Percent | Cumulative Percent |
|--------|----|-----------|---------|---------------|--------------------|
| Valid  | 18 | 7         | .5      | .5            | .5                 |
|        | 19 | 14        | .9      | .9            | 1.4                |
|        | 20 | 19        | 1.3     | 1.3           | 2.7                |
|        | 21 | 17        | 1.1     | 1.1           | 3.8                |
|        | 22 | 29        | 1.9     | 1.9           | 5.8                |
|        | 23 | 13        | .9      | .9            | 6.6                |
|        | 24 | 26        | 1.7     | 1.7           | 8.4                |
|        | 25 | 23        | 1.5     | 1.5           | 9.9                |
|        | 26 | 25        | 1.7     | 1.7           | 11.6               |
|        | 27 | 24        | 1.6     | 1.6           | 13.2               |
|        | 28 | 25        | 1.7     | 1.7           | 14.9               |
|        | 29 | 22        | 1.5     | 1.5           | 16.3               |
|        | 30 | 33        | 2.2     | 2.2           | 18.6               |
|        | 31 | 22        | 1.5     | 1.5           | 20.0               |
|        | 32 | 32        | 2.1     | 2.1           | 22.2               |
|        | 33 | 21        | 1.4     | 1.4           | 23.6               |
|        | 34 | 25        | 1.7     | 1.7           | 25.3               |
|        | 35 | 35        | 2.3     | 2.3           | 27.6               |
|        | 36 | 39        | 2.6     | 2.6           | 30.2               |
|        | 37 | 24        | 1.6     | 1.6           | 31.8               |
|        | 38 | 32        | 2.1     | 2.1           | 34.0               |
|        | 39 | 27        | 1.8     | 1.8           | 35.8               |
|        | 40 | 27        | 1.8     | 1.8           | 37.6               |
|        | 41 | 26        | 1.7     | 1.7           | 39.3               |
|        | 42 | 39        | 2.6     | 2.6           | 41.9               |
|        | 43 | 38        | 2.5     | 2.5           | 44.5               |
|        | 44 | 28        | 1.9     | 1.9           | 46.3               |
|        | 45 | 30        | 2.0     | 2.0           | 48.4               |

| | | | | |
|---|---|---|---|---|
| 46 | 32 | 2.1 | 2.1 | 50.5 |
| 47 | 38 | 2.5 | 2.5 | 53.0 |
| 48 | 41 | 2.7 | 2.7 | 55.8 |
| 49 | 24 | 1.6 | 1.6 | 57.4 |
| 50 | 30 | 2.0 | 2.0 | 59.4 |
| 51 | 30 | 2.0 | 2.0 | 61.4 |
| 52 | 29 | 1.9 | 1.9 | 63.4 |
| 53 | 22 | 1.5 | 1.5 | 64.8 |
| 54 | 37 | 2.5 | 2.5 | 67.3 |
| 55 | 36 | 2.4 | 2.4 | 69.7 |
| 56 | 27 | 1.8 | 1.8 | 71.5 |
| 57 | 11 | .7 | .7 | 72.3 |
| 58 | 24 | 1.6 | 1.6 | 73.9 |
| 59 | 26 | 1.7 | 1.7 | 75.6 |
| 60 | 29 | 1.9 | 1.9 | 77.6 |
| 61 | 15 | 1.0 | 1.0 | 78.6 |
| 62 | 17 | 1.1 | 1.1 | 79.7 |
| 63 | 29 | 1.9 | 1.9 | 81.6 |
| 64 | 22 | 1.5 | 1.5 | 83.1 |
| 65 | 14 | .9 | .9 | 84.1 |
| 66 | 19 | 1.3 | 1.3 | 85.3 |
| 67 | 12 | .8 | .8 | 86.1 |
| 68 | 12 | .8 | .8 | 86.9 |
| 69 | 16 | 1.1 | 1.1 | 88.0 |
| 70 | 11 | .7 | .7 | 88.7 |
| 71 | 17 | 1.1 | 1.1 | 89.9 |
| 72 | 21 | 1.4 | 1.4 | 91.3 |
| 73 | 9 | .6 | .6 | 91.9 |
| 74 | 13 | .9 | .9 | 92.8 |
| 75 | 10 | .7 | .7 | 93.4 |
| 76 | 12 | .8 | .8 | 94.2 |
| 77 | 7 | .5 | .5 | 94.7 |
| 78 | 4 | .3 | .3 | 95.0 |
| 79 | 10 | .7 | .7 | 95.6 |
| 80 | 7 | .5 | .5 | 96.1 |
| 81 | 11 | .7 | .7 | 96.9 |
| 82 | 6 | .4 | .4 | 97.3 |
| 83 | 5 | .3 | .3 | 97.6 |
| 84 | 4 | .3 | .3 | 97.9 |
| 85 | 3 | .2 | .2 | 98.1 |
| 86 | 8 | .5 | .5 | 98.6 |
| 87 | 2 | .1 | .1 | 98.7 |
| 88 | 6 | .4 | .4 | 99.1 |
| 89 OR OLDER | 13 | .9 | .9 | 100.0 |
| Total | 1493 | 99.5 | 100.0 | |
| Missing NA | 7 | .5 | | |
| Total | 1500 | 100.0 | | |

**AGE4: AGE RECODED INTO 4 CATEGORIES**

|  |  | Frequency | Percent | Valid Percent | Cumulative Percent |
|---|---|---|---|---|---|
| Valid | 18 – 34 | 377 | 25.1 | 25.3 | 25.3 |
|  | 35 – 49 | 480 | 32.0 | 32.2 | 57.4 |
|  | 50 – 64 | 384 | 25.6 | 25.7 | 83.1 |
|  | 65+ | 252 | 16.8 | 16.9 | 100.0 |
|  | Total | 1493 | 99.5 | 100.0 |  |
| Missing | System | 7 | .5 |  |  |
| Total |  | 1500 | 100.0 |  |  |

6.

**DEGREE: RS HIGHEST DEGREE**

|  |  | Frequency | Percent | Valid Percent | Cumulative Percent |
|---|---|---|---|---|---|
| Valid | LT HIGH SCHOOL | 222 | 14.8 | 14.8 | 14.8 |
|  | HIGH SCHOOL | 768 | 51.2 | 51.2 | 66.0 |
|  | JUNIOR COLLEGE | 114 | 7.6 | 7.6 | 73.6 |
|  | BACHELOR | 262 | 17.5 | 17.5 | 91.1 |
|  | GRADUATE | 133 | 8.9 | 8.9 | 100.0 |
|  | Total | 1499 | 99.9 | 100.0 |  |
| Missing | NA | 1 | .1 |  |  |
| Total |  | 1500 | 100.0 |  |  |

**DEGREE2: Rs HIGHEST DEGREE (2 CATEGORY RECODE)**

|  |  | Frequency | Percent | Valid Percent | Cumulative Percent |
|---|---|---|---|---|---|
| Valid | Less than College | 1104 | 73.6 | 73.6 | 73.6 |
|  | College Graduate | 395 | 26.3 | 26.4 | 100.0 |
|  | Total | 1499 | 99.9 | 100.0 |  |
| Missing | System | 1 | .1 |  |  |
| Total |  | 1500 | 100.0 |  |  |

The recoded version of DEGREE2 highlights the relative rarity of a college education. Just 26.4 percent of the sample has completed a four year college degree. And, looking at the distribution for DEGREE, we find that over one-third of those with at least a college degree (8.9 percent) have a graduate degree.

# CHAPTER THREE
## Charts and Graphs

_Learning Objectives: By the end of this chapter, you will be able to:_

1. _Construct and analyze bar and pie charts, histograms, line graphs, and population pyramids._

## KEY TERMS

- ✓ Bar chart  (p. 61)
- ✓ Frequency polygon  (p. 65)
- ✓ Histogram  (p. 63)
- ✓ Line chart  (p. 65)
- ✓ Pie chart  (p. 59)
- ✓ Population pyramid  (p. 67)

## CHAPTER SUMMARY

After you display your data in the form of frequency distribution tables (see chapter 2), a second way to display your data is with charts and graphs.  Researchers display their data in charts and graphs because they are visually more appealing and dramatic.  Charts and graphs are also useful visualizations of the shape of a distribution.  Like everything else, the chart or graph you use will depend largely on the level of measurement of your variable.  For nominal variables (and for all variables that have fewer categories), you should use pie charts and bar graphs.  For ordinal and interval-ratio variables (or variables that have many values), you should use histograms and line charts.  A fifth kind of graph, population pyramids, are used for the demographics (often sex and age) of very large groups, societies, or countries.

## CHAPTER OUTLINE

### 1.1  Graphs for Nominal-Level Variables.
- Pie Charts.
  - Pie charts are easy to construct.  Make a circle and divide the circle into slices, as you would on a pie, with each slice corresponding to the proportion (or percentage) of respondents in each category.
  - When a nominal variable has more than five or six categories, bar charts are preferred over pie charts.

- Bar Charts.
  - Bar charts use vertical bars to represent the frequency of cases in each category. Each category is labeled along the horizontal axis on the bottom, and either the frequencies or percentages are marked along the vertical axis on the left-hand side. For each category, make a vertical bar that reaches to the frequency for that category. The categories for nominal and ordinal variables are discrete; therefore the bars of each category should not touch each other.
  - "Grouped" bar charts are especially useful when you want to emphasize the comparison of specific categories.

## 1.2 Graphs for Interval-Ratio-Level Variables.

- Histograms.
  - For histograms, you set the chart up the same way as you would a bar chart, with the frequencies along the vertical axis and the class intervals (or scores) along the horizontal axis. You want your frequency bars to touch each other in a histogram, starting and ending at the real limits of the class intervals.
  - Histograms are used for interval-ratio variables, particularly those with many values.
  - An important point to remember about histograms (and bar charts) is that the higher the bar, the more frequent the value.
- Line Charts.
  - Similarly, line charts (sometimes called frequency polygons) graphically display the distribution of interval-ratio data with a line that connects the frequency at each class interval's midpoint. Like bar charts and histograms, line charts display frequencies along the vertical axis, but they display midpoints of the class intervals along the horizontal axis. The frequency of each class interval is marked with a dot at the midpoint, and then a straight line is used to connect the dots.
  - This kind of chart graphically illustrates the distribution of scores by categories; it is very useful when examining trends over time.
  - Researchers can use multiple lines and/or multiple axes to represent multiple distributions at a time – an extremely useful tool when making comparisons.

## 1.3 Population Pyramids.

- Used to represent demographic characteristics of whole groups or societies, population pyramids are common and efficient graphing tools.
- Population pyramids have two axes. The horizontal axis represents numbers or percentages of people/cases, although percentages are usually preferred. The vertical axis represents ages, most often in 5 year intervals. Men tend to be represented on the left side of the pyramid, and women on the right side. While pyramids may present gender differences most often, the two sides of the pyramid can be constructed to highlight the comparison of some other two groups, like whites and blacks or like the unemployed and employed.

## MULTIPLE-CHOICE QUESTIONS

1.  To graphically display the distribution of a nominal variable, you should use a:
    a. pie chart or bar chart
    b. population pyramid
    c. bar chart or line chart
    d. histogram or line chart

    *Learning Objective: 1*

2.  To graphically display the distribution of an ordinal variable, you should use a:
    a. pie chart or bar chart
    b. population pyramid
    c. bar chart or line chart
    d. histogram or line chart

    *Learning Objective: 1*

3.  To graphically display the distribution of an interval-ratio variable, you should use a:
    a. pie chart or bar chart
    b. population pyramid
    c. bar chart or line chart
    d. histogram or line chart

    *Learning Objective: 1*

4.  In a _____ the size of the segments represents the proportion or percentage of cases in the category.
    a. bar chart
    b. pie chart
    c. line chart
    d. population pyramid

    *Learning Objective: 1*

5.   In a line chart, frequencies are graphed with the use of
     _____.
     a. vertical bars
     b. horizontal bars
     c. real class limits
     d. midpoints

     *Learning Objective: 1*


6.   In a histogram, frequencies are graphed with the use of _____.
     a. vertical bars
     b. horizontal bars
     c. real class limits
     d. midpoints

     *Learning Objective: 1*


7.   Pie charts should be used for variables with _____.
     a. class intervals
     b. midpoints
     c. fewer than six categories
     d. more than six categories

     *Learning Objective: 1*


8.   To graphically display the distribution of an interval-ratio variable by categories of a
     nominal variable.
     a. pie chart or bar chart
     b. population pyramid
     c. bar chart or line chart
     d. histogram or line chart

     *Learning Objective: 1*


9.   The larger the slice in a pie chart, the _____.
     a. greater the relative size of that category
     b. greater the change in the size of that category over time
     c. smaller the relative size of that category
     d. smaller the change in the size of that category over time

     *Learning Objective: 1*

10. In order to construct a pie chart, you must first _____.
    a. calculate the percentage for each category
    b. calculate the percentage change for each category
    c. draw horizontal and vertical axes
    d. calculate real class limits and midpoints

    *Learning Objective: 1*


11. The taller the bar in a bar chart, the _____.
    a. smaller the relative size of that category
    b. smaller the change in the size of that category over time
    c. greater the relative size of that category
    d. greater the change in the size of that category over time

    *Learning Objective: 1*


12. In bar charts and histograms, _____ appear on the vertical axis.
    a. titles
    b. categories
    c. slices
    d. frequencies

    *Learning Objective: 1*


13. In bar charts and histograms, _____ appear on the horizontal axis.
    a. titles
    b. categories
    c. slices
    d. frequencies

    *Learning Objective: 1*


14. In _____ the sides of the bars touch, while in _____ the sides of the bars do not touch.
    a. line chart, bar chart
    b. histogram, pie chart
    c. histogram, bar chart
    d. bar chart, histogram

    *Learning Objective: 1*

15. Researchers can use _____ to compare the relative frequencies for two or more categories of a variable.
    a. population pyramids
    b. grouped bar charts
    c. pie charts
    d. histograms

    *Learning Objective: 1*


16. _____ are especially useful in displaying change over time.
    a. line charts
    b. histograms
    c. bar charts
    d. pie charts

    *Learning Objective: 1*


17. Line charts use _____ instead of _____ to represent the frequency of each category/value.
    a. bars, lines
    b. bars, dots
    c. slices, bars
    d. dots, bars

    *Learning Objective: 1*


18. More than one distribution can be simultaneously displayed in a line chart through the use of _____.
    a. grouped bars
    b. multiple axes
    c. multiple charts
    d. pyramid sides

    *Learning Objective: 1*


19. In population pyramids, _____ appear on the horizontal axis.
    a. sexes
    b. age categories
    c. slices
    d. frequencies

    *Learning Objective: 1*

20.    In population pyramids, _____ often appear on the vertical axis.
       a.  sexes
       b.  age categories
       c.  slices
       d.  frequencies

*Learning Objective: 1*


## WORK PROBLEMS

1.    A researcher collected data on attitudes toward abortion.  The first step in analyzing the
      data was to make the frequency distribution table below (from chapter 2).  Graphically
      display the distribution using a pie chart.

| Attitude Toward Abortion | Frequency (f) |
|---|---|
| Strongly favor | 42 |
| Favor | 57 |
| Oppose | 52 |
| Strongly oppose | 43 |
|  | N = 194 |

*Learning Objective: 1*


2.    Using the data from problem 1, graphically display the distribution using a bar chart.

*Learning Objective: 1*


3.    Using the data from problem 1, graphically display the distribution using a line chart.  Of
      the three charts (pie chart, bar chart, and line chart), which do you think better represents
      the distribution and why?

*Learning Objective: 1*


4.    The data below represent the age at which a sample of newlyweds got married.
      Graphically display the data using a line chart.

          17   25   27   22   35   23   27   28   19   19   22
          18   18   24   31   16   18   26   21   21   23   36
          24   19   22   27   21   23   25   22   22   20

*Learning Objective: 1*

5. The data for age first married was broken down by gender. Graphically display the data using a multiple-line line chart.

| Males | | Females | |
|-------|-----|---------|-----|
| 24 | 27 | 17 | 18 |
| 36 | 23 | 18 | 23 |
| 21 | 19 | 22 | 19 |
| 25 | 26 | 21 | 27 |
| 19 | 25 | 35 | 22 |
| 23 | 22 | 16 | 21 |
| 27 | 28 | 18 | 22 |
| 24 | 31 | 19 | 20 |

*Learning Objective: 1*


6. Using the data from problem 4, construct a histogram for the distribution.

*Learning Objective: 1*


7. Using the data from problem 5, construct two histograms for the distributions.

*Learning Objective: 1*


8. The following questions were asked of 20 respondents.

Sex:   1 Male
       2 Female

Education (given in actual years)

Should marijuana be legalized?
       1 Yes
       2 No

The data for these questions appear on the next page. Graphically display each of the three variables.

| Case | Sex | Education | Legalize Marijuana |
|------|-----|-----------|--------------------|
| 1 | 1 | 12 | 1 |
| 2 | 2 | 12 | 2 |
| 3 | 1 | 14 | 2 |
| 4 | 1 | 15 | 1 |
| 5 | 2 | 11 | 1 |
| 6 | 2 | 9 | 2 |
| 7 | 1 | 16 | 1 |
| 8 | 1 | 16 | 1 |
| 9 | 2 | 14 | 2 |
| 10 | 2 | 12 | 1 |
| 11 | 2 | 11 | 1 |
| 12 | 1 | 12 | 1 |
| 13 | 1 | 12 | 2 |
| 14 | 1 | 17 | 1 |
| 15 | 2 | 18 | 2 |
| 16 | 1 | 10 | 2 |
| 17 | 1 | 14 | 1 |
| 18 | 2 | 15 | 2 |
| 19 | 1 | 9 | 2 |
| 20 | 1 | 11 | 1 |

*Learning Objective: 1*

9.     Re-graph the three distributions in problem 7, using a different (but still appropriate) graph type for each than you already used.

*Learning Objective: 1*

10.    Construct a bar chart with the following data.

| Years of Education | Frequency (f) |
|--------------------|---------------|
| 7 – 9 | 19 |
| 10 – 12 | 57 |
| 13 – 15 | 36 |
| 16 – 18 | 22 |
| 19 – 21 | 10 |
| | N = 144 |

*Learning Objective: 1*

11. Graphically display the following data using a pie chart.

| Highest Educational Degree | Frequency (f) |
|---|---|
| Less than high school | 8 |
| High school | 27 |
| Associates or junior college | 15 |
| Bachelors | 12 |
| Graduate | 3 |
| | N = 65 |

*Learning Objective: 1*

12. The following data about drinking patterns were collected at a large university. Make a multiple-bar bar chart which includes all three majors by the amount of drinking per week.

| Frequency of Drinking | Social Sciences | Major Arts and Humanities | Physical Sciences |
|---|---|---|---|
| Less than once/week | 12 | 8 | 6 |
| Once or twice/week | 17 | 5 | 8 |
| Three to five times/week | 13 | 4 | 6 |
| Five to seven times/week | 6 | 2 | 3 |
| More than seven times/week | 2 | 0 | 2 |
| | N = 50 | N = 19 | N = 25 |

*Learning Objective: 1*

13. Construct a multiple-line line chart for the GPA data (below).

| Freshman | | | | | | Seniors | | | | |
|---|---|---|---|---|---|---|---|---|---|---|
| 2.50 | 3.64 | 3.11 | 4.00 | 0.25 | | 2.05 | 3.44 | 4.00 | 3.43 | 1.90 |
| 3.36 | 3.14 | 2.00 | 2.00 | 1.25 | | 3.56 | 2.11 | 3.88 | 3.50 | 2.33 |
| 4.00 | 2.78 | 1.89 | 3.87 | 3.50 | | 4.00 | 3.25 | 2.71 | 2.84 | 2.00 |
| 0.75 | 2.91 | 3.98 | 2.25 | 2.56 | | 3.11 | 2.66 | 3.00 | 3.19 | 2.50 |
| 3.21 | 4.00 | 3.43 | 3.64 | 1.75 | | 2.75 | 3.46 | 3.13 | 2.05 | 2.75 |
| 1.99 | 1.05 | 2.33 | 1.95 | 4.00 | | 3.55 | 1.95 | 2.25 | 2.68 | 3.68 |
| 2.75 | 2.47 | 3.66 | 3.00 | 3.10 | | 2.45 | 2.00 | 2.97 | 3.47 | 1.95 |
| 3.68 | 3.50 | 4.00 | 2.68 | 4.00 | | 2.95 | 2.19 | 2.68 | 3.90 | 2.44 |
| 3.33 | 2.75 | 2.75 | 3.75 | 1.00 | | 2.76 | 3.52 | 3.12 | 2.49 | 3.56 |
| 2.86 | 3.24 | 2.50 | 0.50 | 2.25 | | 3.48 | 2.58 | 2.94 | 3.00 | 1.75 |

*Learning Objective: 1*

14. Recode the GPAs from problem 13 into categories, using class intervals of 0.24 (for example: 0-0.24, 0.25-0.49, 0.50-0.74, ... , 3.25-3.49, 3.50-3.74, 3.75-4.00). Construct a multiple-bar bar chart for the GPA distributions.

*Learning Objective: 1*

15. Using the recoded GPA data from problem 14, construct a population pyramid where freshman GPAs are on the left and senior GPAs are on the right.

*Learning Objective: 1*

**SPSS or MicroCase WORK PROBLEMS**

The following work problems can be completed using either SPSS or MicroCase. The solutions will be the same regardless of which one you use. Your instructor may have a preference, so be sure to check with him/her prior to beginning these problems.

1. Generate a pie chart for *coneduc*.

2. Generate a bar chart for *childs*.

3. Generate a line chart for *tvhours*.

4. Generate a multiple-bar bar chart for *coneduc* by *sex*.

5. Generate a multiple-line line chart for *tvhours* by *sex*.

6. Generate all appropriate charts for the variable *sei*.

## ANSWER KEY

### MULTIPLE-CHOICE QUESTIONS

1.  a  (p. 59-61)
2.  d  (p. 63-65)
3.  d  (p. 63-65)
4.  b  (p. 59)
5.  d  (p. 65)
6.  a  (p. 63)
7.  c  (p. 59-61)
8.  b  (p. 67)
9.  a  (p. 59-61)
10. a  (p. 59-61)

11. c  (p. 61)
12. d  (p. 61, 63)
13. b  (p. 61, 63)
14. c  (p. 61, 63)
15. b  (p. 62-63)
16. a  (p. 65)
17. d  (p. 65)
18. b  (p. 65-67)
19. d  (p. 67-69)
20. b  (p. 67-69)

### WORK PROBLEMS

1.

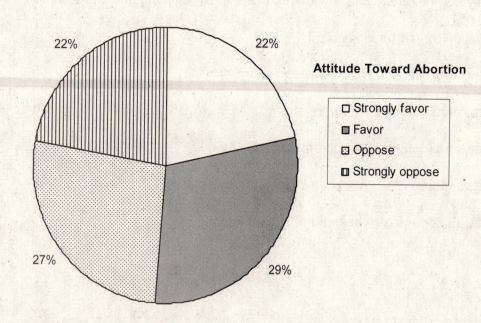

**Attitude Toward Abortion**

☐ Strongly favor
◼ Favor
☒ Oppose
▥ Strongly oppose

2.

**Attitude Toward Abortion**

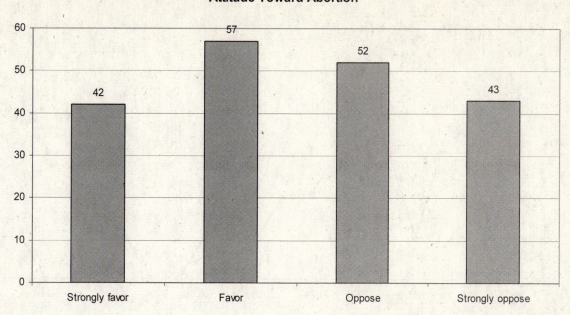

3.

**Attitude Toward Abortion**

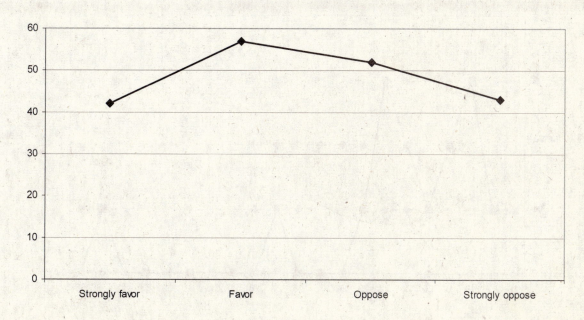

4.

**Age at Marriage**

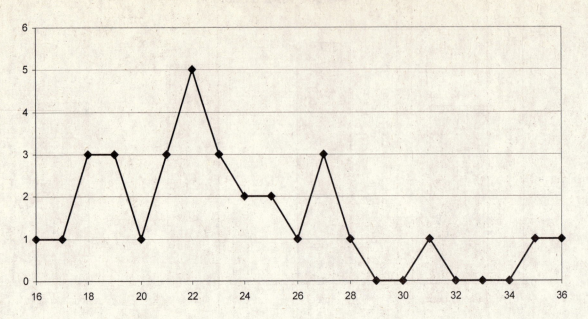

5.

**Age at Marriage**

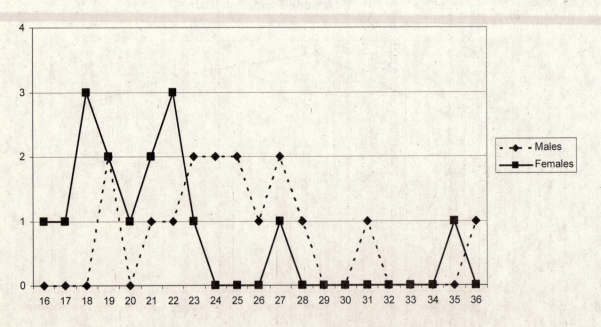

6.

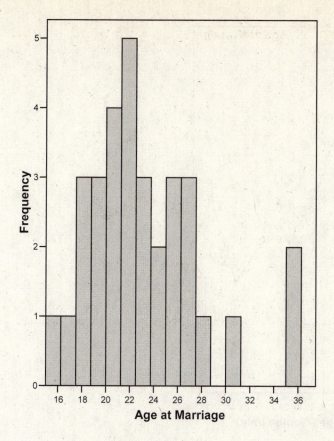

7.

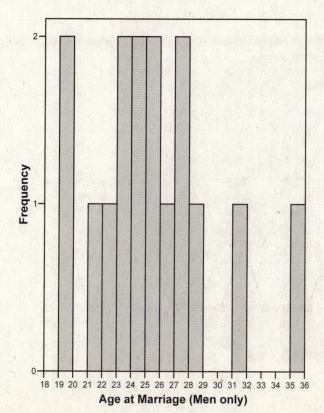

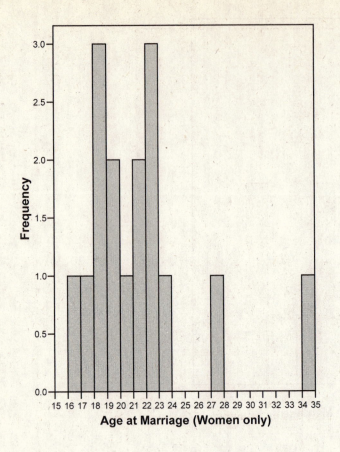

 8.

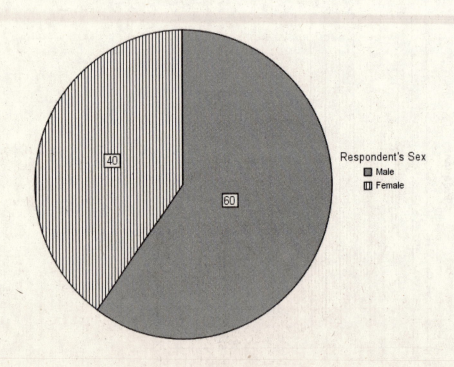

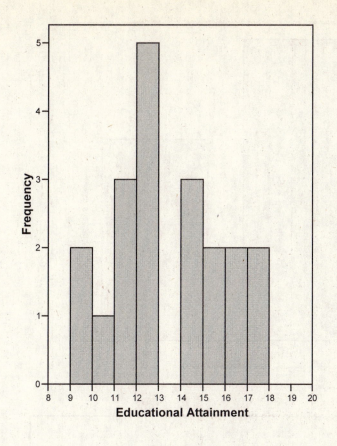

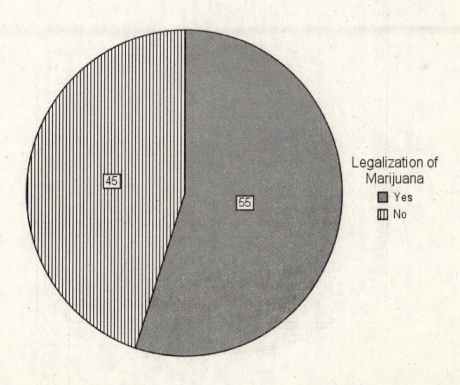

9.

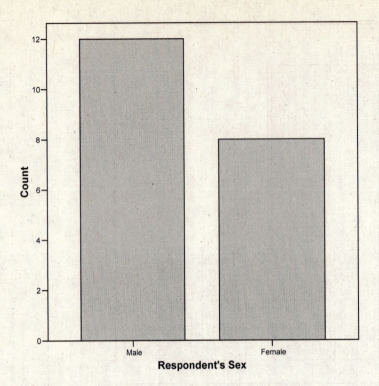

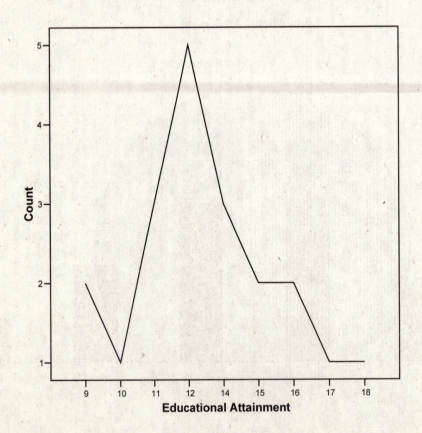

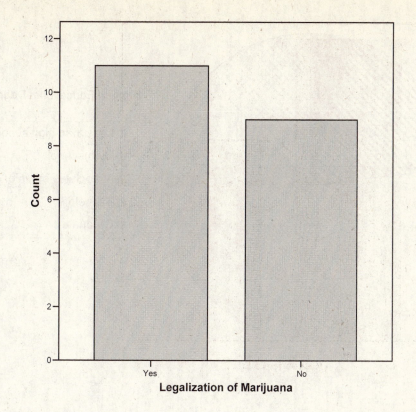

**Legalization of Marijuana**

10.

Years of Education

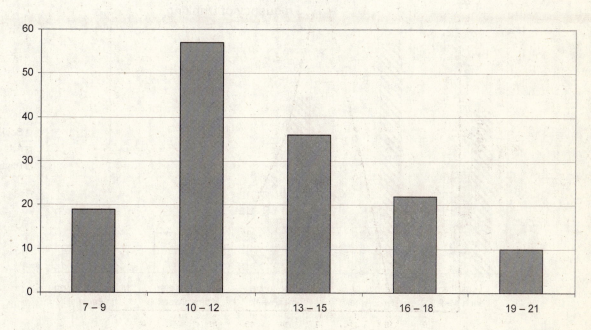

11.

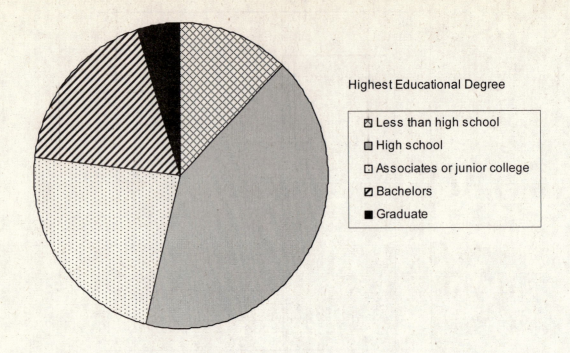

Highest Educational Degree

| | |
|---|---|
| ▨ | Less than high school |
| ▨ | High school |
| ▨ | Associates or junior college |
| ▨ | Bachelors |
| ■ | Graduate |

12.

**Frequency of Drinking**

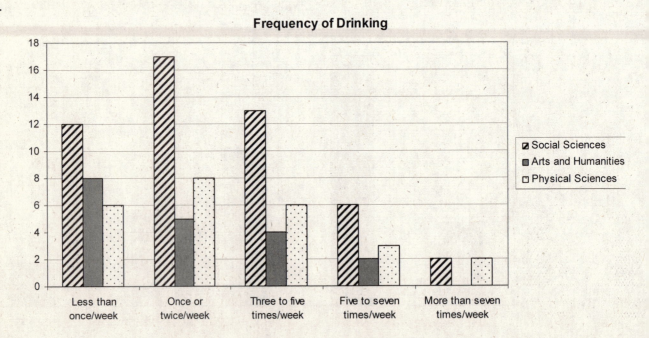

**13.**

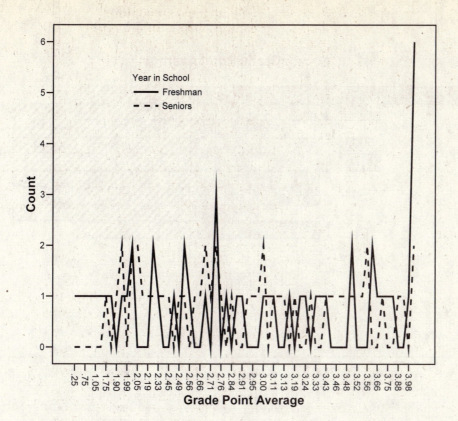

**14.**

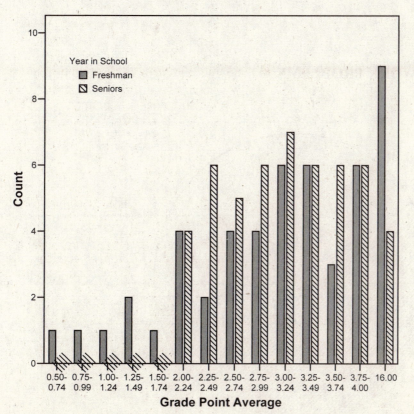

15.

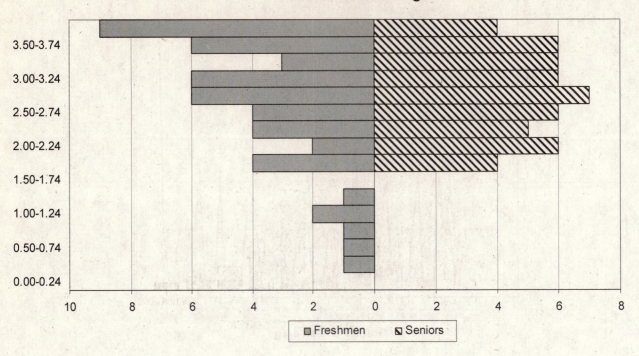

**Grade Point Average**

**SPSS or MicroCase WORK PROBLEMS**

1.

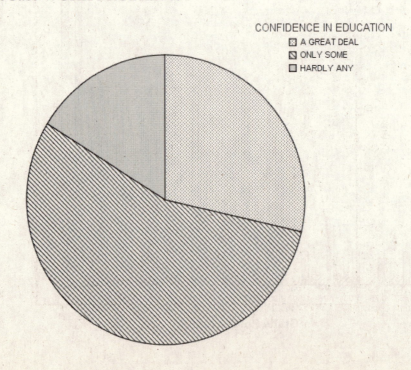

CONFIDENCE IN EDUCATION
- A GREAT DEAL
- ONLY SOME
- HARDLY ANY

2.

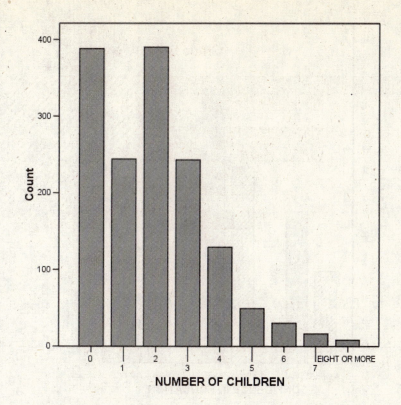

3.

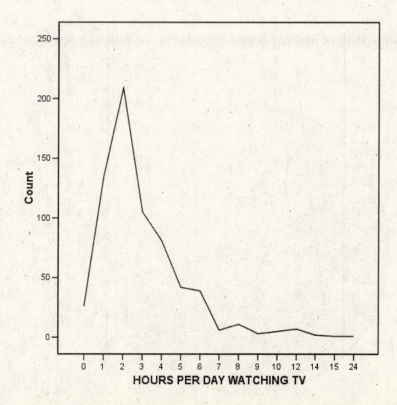

4.

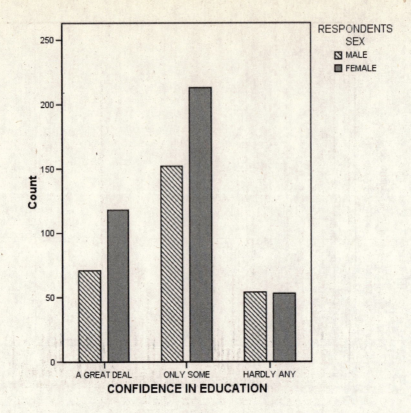

5.

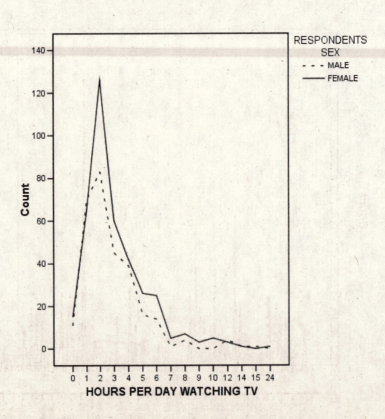

6.

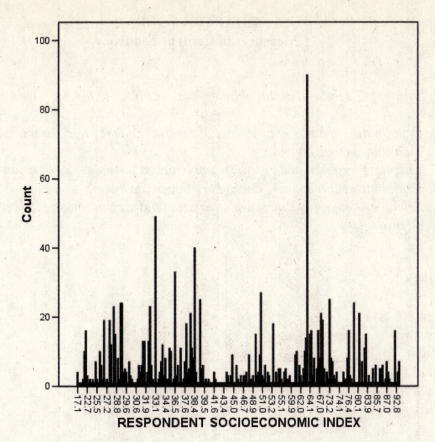

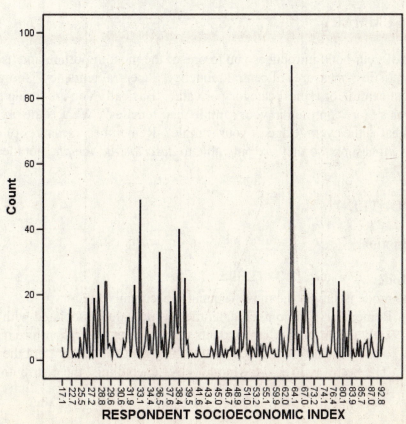

# CHAPTER FOUR
## Measures of Central Tendency

_Learning Objectives_: _By the end of this chapter, you will be able to:_

1.  Explain the purposes of measures of central tendency and the interpret the information they convey.
2.  Calculate, explain, and compare and contrast the mode, median, and mean.
3.  Explain the mathematical characteristics of the mean.
4.  Select an appropriate measure of central tendency according to level of measurement.

## KEY TERMS

- ✓ Mean  (p. 85)
- ✓ Measures of central tendency  (p. 85)
- ✓ Median  (p. 85)
- ✓ Mode  (p. 85)
- ✓ Skew  (p. 93)
- ✓ $X_i$  (p. 89)

## CHAPTER SUMMARY

Chapter four of your book introduces you to one of the most important and useful types of descriptive statistics:  measures of central tendency.  "Central tendency" simply means the most common, most central, or typical category or value.  Basically you are trying to find the average case:  Are males more common in your sample than females?  What is the predominant religious category?  What is the average age of your sample?  Remember: measures of central tendencies are descriptive measures, so they are only able to _describe the sample_, not the population.

## CHAPTER OUTLINE

### 4.1    Introduction.

### 4.2    The Mode.
- The mode is simply the most frequently occurring category.
    - Remember:  for nominal data, there is no <u>value</u> associated with each category, only the number of respondents in each category.  Therefore if there are 56 males and 47 females, then the mode is the category "males", not the frequency of 56. The best way to find the mode is by "eyeballing" the distribution; simply find the category with the greatest number of cases.

- In some distributions, there will not be only <u>one</u> mode, but perhaps two or more modes, and some distributions will not have a mode at all. If this is the case, then report it as such. State that there is not one most frequently occurring category, but several.
- When it comes to ordinal or interval-ratio variables, the mode may fall far away from the true center of the distribution (as indicated by the median and mean).

## 4.3 The Median.

- The median is simply the exact <u>center</u> of the distribution. Unlike the mode, it is affected by the other scores in the distribution, but it is not sensitive to extreme values in the distribution the way the mean is.
- To calculate the median you simply find the middle-most score: order the scores and find the one in the middle. If you have an odd number of cases, add one to N and then divide by two. For example, if you have 31 respondents in your sample, the $16^{th}$ case would be in the exact center of your distribution, with half of the cases falling above and half falling below $[(31+1)/2=16]$. If you have an even number of cases, find the two middle-most scores [N/2 <u>and</u> 1+N/2]. In Table 3.1 below, the two middle-most scores are the $10^{th}$ [20/2=10] and the $11^{th}$ [1+20/2=11] cases, both of which are in the "Juniors" category.

**TABLE 4.1   CLASS RANKINGS IN SOCIOLOGY STATISTICS COURSE**

| Class Ranking | Frequency | Cumulative Frequency | Percentage |
|---|---|---|---|
| Freshmen | 3 | 3 | 15 |
| Sophomores | 5 | 8 | 25 |
| Juniors | 8 | 16 | 40 |
| Seniors | 4 | 20 | 20 |
|  | N = 20 |  | 100 % |

- There are times, though, when you might want to calculate the median for interval-ratio data (rather than the usual measure of central tendency, the mean). Usually this is done when there are a few extreme scores which might skew the distribution (such as annual income in the United States). To calculate the median for interval-ratio data, you would find the average of the value of the two middle-most cases.
- The data at the top of the next page represent ages of a study group in a statistics class. There are eight cases, so we want to find the middle of the distribution, or the average of the fourth (8/2=4)and fifth (1+8/2=5) cases.

| X |
|---|
| 18 |
| 18 |
| 19 |
| 19    ← 4th case |
| 20    ← 5th case |
| 20 |
| 20 |
| 21 |

The median then is the average of the values associated with the 4th and 5th cases, or (19 + 20)/2 =19.5. Therefore, the middle-most score is 19.5 years. The median is usually reserved for ordinal level variables, or when there are extreme scores in the distribution of an interval-ratio level variable. Otherwise, the mean is the most common measure of central tendency.

## 4.4   The Mean.
- Primarily used for interval-ratio variables, the mean is the arithmetic <u>average</u> of scores and is calculated by summing all of the scores and dividing by the total number (N) of scores. The mean is the most common measure of central tendency in the social sciences. The mathematical equation is:

$$\overline{X} = \frac{\sum X_i}{N}$$

## 4.5   Three Characteristics of the Mean.
- The mean is the most sensitive measure of central tendency, which is to say that the mean will be affected by extreme scores in the distribution. As such, there are times when the median is reported for interval-ratio data rather than the mean, especially when there are a few extreme cases, such as with income in the U.S. To explain the difference of these measures, try the following exercise:
  - Get a ruler, some tape, and a few pennies. Try to balance the ruler on the tip of your index finger. If you are using a standard 12-inch ruler, the center should be around the six-inch mark. All things being equal, this point is both your mean and your median. It is the center of your distribution (0-12), or your median, and it is the arithmetic average of the scores (sum of 0 to 12 divided by number of scores, 13, = 78/13 = 6).

66

- Now tape a penny to the 12-inch mark and <u>assume</u> that the value has doubled to 24. Now try to balance the ruler. You must move your "fulcrum", or the point around which all of the scores fall, up a bit, probably closer to the 7" mark, because of the added weight of the new value. If you calculate the mean, substituting the extreme score of 24 for 12, it is 6.9. The mean, then, is pulled up by the extreme score. The median, though, does not change. You still have 13 scores, which now range from 0 to 24; the exact center of your distribution is still six. The mean is sensitive to extreme scores and should be used with some caution when a few extreme scores exist in your data.
- This exercise also demonstrates the idea of skewness. If a distribution is normal, then the mean, the median, and the mode are all in about the same place: the exact center of the distribution. If a distribution has a few extremely high scores, the mean is pulled upward, giving the distribution a positive skew. If a distribution has a few extremely low scores, the mean is pulled downward, giving the distribution a negative skew. In the example with the ruler above, when the value 24 is substituted for the value of 12, the mean is pulled upward, giving the distribution a positive skew.
- Another characteristic of the mean is called the "least squares" principle. This characteristic tells us that the mean is the one point in a distribution where the variation of the other scores is at its minimum. The mean is closer to all of the other scores in the distribution than any other measure of central tendency.

## 4.6    Choosing a Measure of Central Tendency.

- Three measures are used to indicate central tendency, and like many other measures, which one you use depends largely on the level of measurement of the variable you are interested in describing.
  - For nominal data, the mode is the only appropriate measure of central tendency.
  - For ordinal data, the mode is appropriate, but the median is the best measure of central tendency; in some situations the mean is used, although it is technically inappropriate.
  - For interval-ratio data, all three measures of central tendency are appropriate, but in most cases the mean gives you the most information. When the distribution of an interval-ratio variable is skewed, though, the median is usually the preferred measure of central tendency.

## MULTIPLE-CHOICE QUESTIONS

1.    Measures of central tendency are _____.
   a. statistics that describe a population
   b. statistics that are used to find deciles
   c. statistics that summarize proportions
   d. statistics that report the most typical case

   *Learning Objective: 1*

2.     Choosing the most appropriate measure of central tendency is based on _____.
       a. sample size
       b. number of categories
       c. number of variables
       d. level of measurement

       *Learning Objective: 4*


3.     For nominal data, the most appropriate measure of central tendency is _____.
       a. median
       b. mode
       c. mean
       d. frequency

       *Learning Objective: 4*


4.     For ordinal data, the most appropriate measure of central tendency is _____.
       a. median
       b. mode
       c. mean
       d. frequency

       *Learning Objective: 4*


5.     For normally distributed interval-ratio data, the most appropriate measure of central
       tendency is _____.
       a. median
       b. mode
       c. mean
       d. frequency

       *Learning Objective: 4*


6.     The median is _____.
       a. the most frequently occurring category
       b. the arithmetic average
       c. the number of cases
       d. the middle-most score

       *Learning Objective: 2*

7. The mean is _____.
   a. the most frequently occurring category
   b. the arithmetic average
   c. the number of cases
   d. the middle-most score

   *Learning Objective: 2*


8. The mode is _____.
   a. the most frequently occurring category
   b. the arithmetic average
   c. the number of cases
   d. the middle-most score

   *Learning Objective: 2*


9. In a distribution of ten scores, the median will be _____.
   a. 5
   b. the $5^{th}$ case
   c. the sum of the $5^{th}$ and $6^{th}$ cases
   d. the average of the $5^{th}$ and $6^{th}$ cases

   *Learning Objective: 2*


10. In a positively skewed distribution there are some extremely _____ scores, which pull the mean _____.
    a. low, downward
    b. high, upward
    c. low, upward
    d. high, downward

    *Learning Objective: 3*


11. The term $X_i$ means _____.
    a. the sum of the frequencies
    b. the sample size
    c. any category in a distribution
    d. any score in a distribution

    *Learning Objective: 2*

12.    For the variable *years married*, which measure of central tendency would be most
       appropriate?
       a. mode
       b. median
       c. mean
       d. none of the above

       *Learning Objective: 4*


13.    For the variable *satisfaction with marriage* (measured as "a great deal," "somewhat," and
       "not at all"), which measure of central tendency would be most appropriate?
       a. mode
       b. median
       c. mean
       d. none of the above

       *Learning Objective: 4*


14.    For the variable *favor inter-racial marriages* (measured as "yes" or "no"), which measure
       of central tendency would be most appropriate?
       a. mode
       b. median
       c. mean
       d. none of the above

       *Learning Objective: 4*


15.    For the variable *age first married* (measured in years), which measure of central tendency
       would be most appropriate?  (Assume a normal distribution.)
       a. mode
       b. median
       c. mean
       d. none of the above

       *Learning Objective: 4*

16. For the variable *income* (measured in dollars and positively skewed), which measure of central tendency would be most appropriate?
    a. mode
    b. median
    c. mean
    d. none of the above

    *Learning Objective: 4*

17. A "quick and easy" indicator of the center of a distribution is _____.
    a. mode
    b. median
    c. mean
    d. none of the above

    *Learning Objective: 2*

18. A limitation of the _____ is that it may be indeterminable for a given distribution.
    a. mode
    b. median
    c. mean
    d. none of the above

    *Learning Objective:*

19. Unlike the _____, there can be multiple _____ in a distribution.
    a. mode, means
    b. mode, medians
    c. mean, medians
    d. mean, modes

    *Learning Objective: 2*

20. The "least squares" principle of the mean can be summarized as _____.
    a. the mean is farther from all other scores in a distribution
    b. the mean is affected by all scores in a distribution
    c. the mean is closer to all other scores in a distribution
    d. the mean acts like a fulcrum in a distribution

    *Learning Objective: 3*

71

**WORK PROBLEMS**

1.    Find the most appropriate measure of central tendency for the frequency distribution below.

| Number of Children in Family | Frequency (f) |
|---|---|
| 0 | 20 |
| 1 | 10 |
| 2 | 56 |
| 3 | 32 |
| 4 | 10 |
| 5 | 2 |
| 6 | 0 |
| 7 or more | 0 |
|  | N = 130 |

*Learning Objective: 2 & 4*

2.    The data below give the ages of 20 students in a sociology statistics course. Find the mode, median, and mean. What does each tell us about the distribution? Is the distribution skewed? Which way?

| 18 | 19 | 20 | 21 | 20 | 19 | 18 | 21 | 19 | 20 |
|---|---|---|---|---|---|---|---|---|---|
| 19 | 18 | 20 | 20 | 20 | 20 | 20 | 21 | 19 | 37 |

*Learning Objective: 2 & 3*

3.    Using the data from problem 2, drop the highest age and then recalculate the mode, median, and mean. How does the distribution change?

*Learning Objective: 2 & 3*

4.    Find the most appropriate measure of central tendency for the frequency distribution below.

| Social Class | Frequency (f) |
| --- | --- |
| Lower | 5 |
| Working | 18 |
| Middle | 37 |
| Upper Middle | 55 |
| Upper | 2 |
| | N = 117 |

*Learning Objective: 2 & 4*

5.    Find the most appropriate measure of central tendency for the frequency distribution below.

| Political Party Affiliation | Frequency (f) |
| --- | --- |
| Democrat | 543 |
| Republican | 469 |
| Other | 80 |
| None | 18 |
| | N = 1100 |

*Learning Objective: 2 & 4*

6.    Find the most appropriate measure of central tendency for the frequency distribution below.

| Class Ranking | Frequency (f) |
| --- | --- |
| Freshmen | 3 |
| Sophomores | 5 |
| Juniors | 15 |
| Seniors | 13 |
| Graduates | 1 |
| | N = 37 |

*Learning Objective: 2 & 4*

7. Find the most appropriate measure of central tendency for the frequency distribution below.

| Grades on Statistics Exam | Frequency (f) |
|---|---|
| 100 | 2 |
| 95 | 12 |
| 85 | 15 |
| 75 | 18 |
| 65 | 15 |
| 55 | 5 |
| | N = 67 |

*Learning Objective: 2 & 4*

8. A-1 College polled their graduates to find out their incomes after graduating from their stellar programs. Below are the results of their poll. Find the mean income.

| Incomes of Recent Graduates of A-1 College | | | | |
|---|---|---|---|---|
| $12,500 | $13,500 | $17,500 | $30,000 | $23,000 |
| $14,350 | $18,900 | $32,450 | $14,000 | $27,650 |
| $33,450 | $26,760 | $19,000 | $21,000 | $37,800 |
| $12,000 | $22,500 | $27,500 | $23,000 | $20,900 |

*Learning Objective: 2*

9. Using the information from problem 8, A-1 College did not include their most prominent alumni, Mike Rosoft, the computer genius whose income was estimated to be $10,000,000. What would happen to the mean if his income was included? What would be a better measure to use if he were included?

*Learning Objective: 2 & 4*

10. Using the following sample of ages of children, demonstrate the "least squares" principle of the mean.

| Ages |
|---|
| 3 |
| 5 |
| 6 |
| 9 |
| 17 |

*Learning Objective: 3*

11. Using the information from problem 10, demonstrate that the mean is the geographic center of any distribution of scores, around which all other scores cancel out.

   *Learning Objective: 3*

12. Calculate all three measures of central tendency for the frequency distribution below. (Numerical codes for each category are provided in the table.) Compare and contrast the three measures, and discuss the appropriateness of each for describing the distribution.

   | Approval of Gay Marriage | Frequency (f) |
   | --- | --- |
   | Strongly disagree (1) | 25 |
   | Disagree (2) | 36 |
   | Neutral (3) | 15 |
   | Agree (4) | 31 |
   | Strongly agree (5) | 18 |
   | | N = 125 |

   *Learning Objective: 2*

13. The data for age first married was broken down by gender. Calculate the means and medians for each of the two distributions. Compare the distributions, making sure to discuss the center and shape of the distributions.

   | Males | | Females | |
   | --- | --- | --- | --- |
   | 24 | 27 | 17 | 18 |
   | 36 | 23 | 18 | 23 |
   | 21 | 19 | 22 | 19 |
   | 25 | 26 | 21 | 27 |
   | 19 | 25 | 35 | 22 |
   | 23 | 22 | 16 | 21 |
   | 27 | 28 | 18 | 22 |
   | 24 | 31 | 19 | 20 |

   *Learning Objective: 2*

14. Find all appropriate measures of central tendency for the frequency distribution below. Discuss why it/they are appropriate.

| Gender | Frequency (f) |
|--------|---------------|
| Men | 500 |
| Women | 500 |
| | N = 1000 |

*Learning Objective: 2 & 4*

## SPSS or MicroCase WORK PROBLEMS

The following work problems can be completed using either SPSS or MicroCase. The solutions will be the same regardless of which one you use. Your instructor may have a preference, so be sure to check with him/her prior to beginning these problems.

1. Find the most appropriate measure of central tendency for each of the following variables: *marital, grass, happy, health, childs*. Discuss the output for each one.

2. Use the Descriptives command to get the means, medians, and modes for each of the following variables: *age, tvhours, realinc, sei*. Characterize the normality or skewness of each variable based on the measures of central tendency.

3. Choose three variables from the GSS (not used in the other problems in this section) and find the most appropriate measure of central tendency for each.

4. Choose an ordinal variable from the GSS (not used in the other problems in this section) and use the Descriptives command to get the mean, median, and mode. Describe situations when you would use each measure.

## ANSWER KEY

### MULTIPLE-CHOICE QUESTIONS

| | | | | | | |
|---|---|---|---|---|---|---|
| 1. | d | (p. 85) | 11. | d | (p. 89) |
| 2. | d | (p. 85) | 12. | c | (p. 90) |
| 3. | b | (p. 86) | 13. | b | (p. 88) |
| 4. | a | (p. 88) | 14. | a | (p. 86) |
| 5. | c | (p. 90) | 15. | c | (p. 90) |
| 6. | d | (p. 85) | 16. | b | (p. 93-94) |
| 7. | b | (p. 85) | 17. | a | (p. 85) |
| 8. | a | (p. 85) | 18. | a | (p. 86) |
| 9. | d | (p. 87-88) | 19. | d | (p. 86) |
| 10. | b | (p. 93) | 20. | c | (p. 91-92) |

### WORK PROBLEMS

1.  The first question to ask is "what is the level of measurement of the variable?"  We cannot find the appropriate measure of central tendency until we know the level of measurement.  The level of measurement for *number of children in family* is interval-ratio, therefore we can use any of the three measures of central tendency, but the mean is usually the best measure for such variables.

    The mean is the sum of scores ($\Sigma X_i$) divided by N, or:

    $$\overline{X} = \frac{(0\times 20) + (1\times 10) + (2\times 56) + (3\times 32) + (4\times 10) + (5\times 2) + (6\times 0) + (7\times 0)}{130}$$

    $$\overline{X} = \frac{268}{130} = 2.06$$

    Therefore, the mean number of children per household is 2.06.

    However, this distribution appears to have a slight positive skew; therefore, the median may be the best measure of central tendency.  With an even number of cases (N=130), the median is the average of the 65th (N/2=130/2) and the 66th (1+N/2=1+130/2) case.  If we add a cumulative frequency column to the frequency distribution, we can easily locate the 65th and 66th cases.  The cumulative frequencies for this table are: 20, 30, 86, 118, 128, 130, 130, and 130.  Thus, the 65th and 66th cases are both in the "2" category, making the median number of children per household two.  You'll notice that the mean is in fact slightly higher than the median, supporting our characterization of the distribution as slightly positively skewed.

2.  Begin by organizing the data from lowest to highest. This makes it easier to see the distribution. The mode, or most frequently occurring case, is 20. Therefore, there are more 20-year olds in this statistics class than any other age.

The median, the exact center of the distribution, is the average of the 10th (N/2=20/2=10) and 11th (1+N/2=1+20/2=11) cases, which are 20 and 20. Therefore, the median is 20 years of age. Remember, you cannot find the median until the data have been ordered.

The mean, or the arithmetic average, is the sum of all scores ($\Sigma X_i$) divided by the total number of respondents (N), or 409/20, which equals 20.45. Therefore, the average age of students in this statistics course is 20.45 years.

Because one student is a bit older than the rest, the mean age is pulled upward; therefore, the distribution has a positive skew.

3.  If you were to drop that case, the mean age would be 19.6 years, which is closer to the mode and median, which are both still 20 years of age, making the distribution more normal in shape.

4.  The level of measurement for the *social class* variable is ordinal, because we can rank-order the classes, but there is not a numerical value associated with each class. Therefore, we can use either the mode or the median, but the median is the best measure of central tendency.

To find the median for a distribution with an odd number of cases (N=117), we need to find the category associated with the 59th case [(N+1)/2=(117+1)/2=59]. The easiest way to do so is to calculate the cumulative frequencies:

| Social Class | Frequency (f) | Cumulative Frequency (cf) | |
|---|---|---|---|
| Lower | 5 | 5 | |
| Working | 18 | 23 | |
| Middle | 37 | 60 | ← the 59th case lies in this category |
| Upper Middle | 55 | 115 | |
| Upper | 2 | 117 | |
| | N = 117 | | |

Therefore, the 59th case falls in the category "middle class." It is very important to remember that the median is not the number 59, but the category in which the 59th case resides.

Make sure to remember the difference in finding the median for "numerical" and "non-numerical" ordinal variables. When we have a "numerical" variable, like the *number of children* variable in problem 2, we take the average of the two middle-most scores. (Because they're "numerical," we can average the two scores.) When we have a "non-numerical" variable, like *social class*, you cannot average the two middle-most scores. Instead, the category in which the two middle-most scores reside is the median.

5. The *political party affiliation* variable is nominal because you cannot rank order the categories. Therefore, the mode is the only appropriate measure of central tendency. The mode is the most frequently occurring category, which is Democrats, with 543 of the respondents falling into this category.

6. The variable *class ranking* is ordinal because the categories can be rank-ordered from highest to lowest. Therefore, either the mode or the median can be used, but the median is the better measure of central tendency.

   Again, calculating the cumulative frequencies will make it easier to find the middle of the distribution. With an odd number of cases (N=37), we're looking for the 19th case $[(N+1)/2 = (37+1)/2 = 19]$.

| Class Ranking | Frequency (f) | Cumulative Frequency (cf) | |
|---|---|---|---|
| Freshmen | 3 | 3 | |
| Sophomores | 5 | 8 | |
| Juniors | 15 | 23 | ← the 19th case lies in this category |
| Seniors | 13 | 36 | |
| Graduates | 1 | 37 | |
| | N = 37 | | |

   The median is the category associated with the 19th case, which is in the category "Juniors." Therefore, half of the distribution falls both above and below the 19th case in the category of Juniors.

7. The level of measurement of the *grades* variable is interval-ratio, which means you can use any of the three measures of central tendency, but the best measure is usually the mean. Because this distribution appears to be normal, the mean is the most appropriate measure of central tendency.

To find the mean, sum all of the scores ($\Sigma X_i$) and divide by the sample size (N). Because the scores are displayed in a frequency table, the easiest way to do is to multiply each value by its frequency, add them together, and divide by N. When you do this, you get:

$$\overline{X} = \frac{\sum X_i}{N} = \frac{(100 \times 2) + (95 \times 12) + (85 \times 15) + (75 \times 18) + (65 \times 15) + (55 \times 5)}{67}$$

$$\overline{X} = \frac{5,215}{67} = 77.84$$

Therefore, the mean grade in this class is 77.8.

8.     *Income* is an interval-ratio variable, so the mean is usually the best measure of central tendency. Often, though, there is such a wide range of incomes that the median is used. In order to determine which measure is best, find both the median and the mean.

To do so, organize the data from lowest to highest. With an even number of cases (N=20) we find the median by taking the <u>average</u> of the 10th (N/2=20/2=10) and 11th (1+N/2=1+20/2=11) cases. Once we organize the data, we find that the 10th case equals $21,000 and the 11th case equals $22,000. Next, we average these two values:

$$Median = \frac{(21000 + 22000)}{2} = \frac{43000}{2} = 21,500$$

The median income, then, is $ 21,500. Half of the alumni make less than $21,500, and half of the alumni make more than $21,500.

The mean is the sum of scores ($\Sigma X_i$) divided by N, or:

$$\overline{X} = \frac{\sum X_i}{N} = \frac{12500 + 17500 + \ldots + 23,000 + 20,900}{20} = \frac{447,760}{20} = 22,388$$

Therefore, the mean income is $22,388. The mean is a bit higher than the median, which indicates a slight positive skew to the distribution. Although the difference is not extreme, it is good idea to report both the median and the mean.

9.     If we include Mike Rosoft's income of $10,000,000 the mean becomes:

$$\overline{X} = \frac{\sum X_i}{N} = \frac{10,447,760}{21} = 497,512.38$$

Therefore, the mean income of the graduating class would be close to $500,000. If you look at the distribution, though, that does not represent the sample at all! Therefore the median would be a better measure of central tendency.

Now that there are 21 cases, the median occurs at the 11th case [(N+1)/2=(21+1)/2=11]. And as we found above for problem 7, the 11th case equals $22,000. Therefore, the median income, once we include Mike Rosoft's value, becomes $22,000. Half of the graduating class make less than $22,000, and half make more than $22,000.

10.   We can demonstrate the "least squares" principle of the mean by showing that the sum of the squared deviations around the mean is lower than the sum of squared deviations around any other value. Recall that a deviation is calculated by subtracting the mean from each individual score ($X_i$).

In order to calculate the deviations, we first need the mean age.

$$\overline{X} = \frac{\sum X_i}{N} = \frac{3+5+6+9+17}{5} = \frac{40}{5} = 8$$

Therefore, the mean age for the five children is eight years.

Now, to demonstrate the "least squares" principle we must take each age, subtract the mean, square that difference, and then sum these squared deviations. Once we've done that, we'll repeat the process, except that we'll use a different value in place of the mean.

| $X_i$ | $(X_i - \overline{X})^2$ | $(X_i - 6)^2$ | $(X_i - 9)^2$ |
|---|---|---|---|
| 3 | $(3-8)^2 = (-5)^2 = 25$ | $(3-6)^2 = (-3)^2 = 9$ | $(3-9)^2 = (-6)^2 = 36$ |
| 5 | $(5-8)^2 = (-3)^2 = 9$ | $(5-6)^2 = (-1)^2 = 1$ | $(5-9)^2 = (-4)^2 = 16$ |
| 6 | $(6-8)^2 = (-2)^2 = 4$ | $(6-6)^2 = (0)^2 = 0$ | $(6-9)^2 = (-3)^2 = 9$ |
| 9 | $(9-8)^2 = (1)^2 = 1$ | $(9-6)^2 = (3)^2 = 9$ | $(9-9)^2 = (0)^2 = 0$ |
| 17 | $(17-8)^2 = (9)^2 = 81$ | $(17-6)^2 = (11)^2 = 121$ | $(17-9)^2 = (8)^2 = 64$ |
| | $\Sigma = 120$ | $\Sigma = 140$ | $\Sigma = 125$ |

The squared deviations appear in the second column of the table. Their sum equals 120, which, according to the "least squares" principle of the mean, is a minimum value. In order to demonstrate this fact, I have selected two other values to substitute for the mean in the equation, 6 and 9. As we can see in the third and fourth columns of the table, when we use any other value the sum of the squared deviations is greater than 120. You can repeat this process with any other number (for example 2, 5, 7.5, etc.), and you will receive the same results.

11. Problem 11 asks us to demonstrate that the mean is the geographic center of the distribution, around which all other scores cancel out. To demonstrate this characteristic of the mean, we only need to show that the sum of the differences between the scores and the mean equals zero.

| $X_i$ | $(X_i - \overline{X})$ |
|-------|------------------------|
| 3 | (3-8) = -5 |
| 5 | (5-8) = -3 |
| 6 | (6-8) = -2 |
| 9 | (9-8) = 1 |
| 17 | (17-8) = 9 |
| | $\Sigma = 0$ |

12. The mode, or most frequently occurring response, is "disagree." Therefore, the most common view is to disagree with gay marriage.

The median, the exact center of the distribution, is the 63rd case, which is "neutral." Therefore, the median or middle-most response is to be neutral on the topic.

The mean, or the arithmetic average, is the sum of all scores ($\Sigma X_i$) divided by the total number of respondents (N). Because this is an ordinal variable, we will have to use the numerical values provided to represent the different responses. We will also have to multiply each of those values by its frequency.

$$\overline{X} = \frac{\sum X_i}{N} = \frac{(1 \times 25) + (2 \times 36) + (3 \times 15) + (4 \times 31) + (5 \times 18)}{125} = 2.85$$

Therefore, the average response is 2.85, which falls between "disagree" and "neutral," but falls closes to the neutral response.

While the median is the most appropriate measure of center for this ordinal variable, we would determine the mode if we just wanted a "quick and easy" measure of center. We might determine the mean, even though it is technically inappropriate, if we intended to use more advanced statistical techniques.

13.     In order to determine the medians for each distribution, the ages must be rearranged in numerical order.

Males:     19, 19, 21, 22, 23, 23, 24, 24, 25, 25, 26, 27, 27, 28, 31, 36

Females:   16, 17, 18, 18, 18, 19, 19, 20, 21, 21, 22, 22, 22, 23, 27, 35

With an even number of cases (N=16) we find the median by taking the average of the 8th (N/2=16/2=8) and 9th (1+N/2=1+16/2=9) cases. Once we organize the data, we find that the 8th male case equals 24 and the 9th male case equals 25. And, we find that the 8th female case equals 20 and the 9th female case equals 21. Next, we average these two values to find that the median age for males is 24.5 and the median age for females is 20.5.

The mean, or the arithmetic average, is the sum of all scores ($\Sigma X_i$) divided by the total number of respondents (N). For males: $\Sigma X_i /N = 400/16 = 25$. For females: $\Sigma X_i /N = 338/16 = 21.13$. Therefore, the average age is 25 for males and 21.13 for females.

14.     The *gender* variable is nominal because you cannot rank order the categories. Therefore, the mode is the only appropriate measure of central tendency. However, because the sample is evenly split across the two gender categories, this distribution does not have a mode.

## SPSS or MicroCase WORK PROBLEMS

If you had problems with any of these questions, be sure to re-read the appropriate sections at the end of chapter 4 in your textbook.

1.      I make it easier on myself by analyzing the variables together based on the level of measurement of each. Thus, I lump all the nominal variables together, then the ordinal ones, then the interval-ratio ones. Because I want different statistics for each level of measurement, it makes the analysis go much faster.

The variable *grass* is a nominal-level variable, so I'm interested in the mode (or most frequently occurring category), which is the "not legal" category. Thus the majority of the sample (63.5%) stated that they did not believe marijuana should be made legal.

**GRASS: SHOULD MARIJUANA BE MADE LEGAL**

| | | Frequency | Percent | Valid Percent | Cumulative Percent |
|---|---|---|---|---|---|
| Valid | LEGAL | 226 | 15.1 | 36.5 | 36.5 |
| | NOT LEGAL | 393 | 26.2 | 63.5 | 100.0 |
| | Total | 619 | 41.3 | 100.0 | |
| Missing | NAP | 833 | 55.5 | | |
| | NA | 48 | 3.2 | | |
| | Total | 881 | 58.7 | | |
| Total | | 1500 | 100.0 | | |

The variable *marital* is also a nominal-level variable. So, again, I'm interested in the mode, which is the "married" category. Therefore, a majority of the sample (48.3%) is married.

**MARITAL: MARITAL STATUS**

| | | Frequency | Percent | Valid Percent | Cumulative Percent |
|---|---|---|---|---|---|
| Valid | MARRIED | 724 | 48.3 | 48.3 | 48.3 |
| | WIDOWED | 130 | 8.7 | 8.7 | 57.0 |
| | DIVORC-SEPARATED | 287 | 19.1 | 19.2 | 76.2 |
| | NEVER MARRIED | 357 | 23.8 | 23.8 | 100.0 |
| | Total | 1498 | 99.9 | 100.0 | |
| Missing | 9 | 2 | .1 | | |
| Total | | 1500 | 100.0 | | |

Because both general happiness (*happy*) and health condition (*health*) are ordinal-level variables, I ran their frequency distributions at the same time. The median for general happiness is category 2, or "Pretty Happy." I found the median by looking at the cumulative frequency column, where I can see that 50% falls in the second category, "Pretty Happy." Thus, we know that the exact center of the distribution of the sample falls within this category (cumulative percent: 85.6%). The median for health condition also occurs in the second category, "good." The fifty percent cumulative percent falls in this category (cumulative percent: 76.8%).

**HAPPY: GENERAL HAPPINESS**

| | | Frequency | Percent | Valid Percent | Cumulative Percent |
|---|---|---|---|---|---|
| Valid | VERY HAPPY | 329 | 21.9 | 32.9 | 32.9 |
| | PRETTY HAPPY | 526 | 35.1 | 52.7 | 85.6 |
| | NOT TOO HAPPY | 144 | 9.6 | 14.4 | 100.0 |
| | Total | 999 | 66.6 | 100.0 | |
| Missing | NAP | 497 | 33.1 | | |
| | NA | 4 | .3 | | |
| | Total | 501 | 33.4 | | |
| Total | | 1500 | 100.0 | | |

**HEALTH : CONDITION OF HEALTH**

| | | Frequency | Percent | Valid Percent | Cumulative Percent |
|---|---|---|---|---|---|
| Valid | EXCELLENT | 325 | 21.7 | 28.0 | 28.0 |
| | GOOD | 566 | 37.7 | 48.8 | 76.8 |
| | FAIR | 211 | 14.1 | 18.2 | 95.0 |
| | POOR | 58 | 3.9 | 5.0 | 100.0 |
| | Total | 1160 | 77.3 | 100.0 | |
| Missing | NAP | 339 | 22.6 | | |
| | NA | 1 | .1 | | |
| | Total | 340 | 22.7 | | |
| Total | | 1500 | 100.0 | | |

For the interval-ratio variable number of children (*childs*), I used the descriptive statistics command. This gave me the output needed, mean number of children. The mean number of children for the sample is 1.92.

**Descriptive Statistics**

| | N | Minimum | Maximum | Mean | Std. Deviation |
|---|---|---|---|---|---|
| NUMBER OF CHILDREN | 1497 | 0 | 8 | 1.92 | 1.666 |

2.    In order to save space and time, I asked SPSS for the means, medians, and modes for all of the variables simultaneously, using the "Statistics" option in the "Frequencies" window.

**Statistics**

| | | AGE: AGE OF RESPONDENT | TVHOURS: HOURS PER DAY WATCHING TV | REALINC: FAMILY INCOME IN CONSTANT $ | SEI: RESPONDENT SOCIOECONOMIC INDEX |
|---|---|---|---|---|---|
| N | Valid | 1493 | 673 | 1282 | 1426 |
| | Missing | 7 | 827 | 218 | 74 |
| Mean | | 47.43 | 2.96 | 33049.90 | 50.017 |
| Median | | 46.00 | 2.00 | 24782.00 | 44.700 |
| Mode | | 48 | 2 | 37173 | 63.5 |

According to the output above, the mean age is 47.43 years; the median is 46 years; and the mode is 48 years. Because the mean is larger than the other two values, that indicates that there are some extreme high values in the distribution. Therefore, the age distribution is positively skewed.

The mean income for the sample is $33,049.90; the median income is $24,782.00; and the modal income is $37,173. Because the mean is greater than the median, there are some extreme high scores in the distribution. Therefore, like age, the income distribution is positively skewed.

The mean SEI value is 50.017; the median value is 44.700; and the mode value is 63.5. Again, because the mean is greater than the median, there are some extreme high scores in the distribution, and the SEI distribution is positively skewed.

As with the other three distributions, the mean number of hours is 2.96, and the median number is 2; this makes the TV hours variable positively skewed too.

# CHAPTER FIVE
## Measures of Dispersion

_Learning Objectives_: By the end of this chapter, you will be able to:

1.  Explain the purpose of measures of dispersion and the information they convey.
2.  Compute and explain the range (R), the interquartile range (Q), the standard deviation (s), and the variance ($s^2$).
3.  Select an appropriate measure of dispersion and correctly calculate and interpret the statistic.
4.  Describe and explain the mathematical characteristics of the standard deviation.


## KEY TERMS

- ✓ Deviations  (p. 109)
- ✓ Dispersion  (p. 105)
- ✓ Interquartile range  (p. 107)
- ✓ Measures of dispersion  (p. 105)
- ✓ Range  (p. 106)
- ✓ Standard deviation  (p. 110)
- ✓ Variance  (p. 110)


## CHAPTER SUMMARY

Measures of dispersion are another way of describing a distribution of scores on some given variable. Like measures of central tendencies, measures of dispersion give the reader a sense of what the sample looks like, but these measures add a bit more insight. We know what the average score looks like with either a mode, median, or mean, but with these new measures we can go a step further and ask "How much variation is there in the data?"


## CHAPTER OUTLINE

### 5.1  Introduction.
- If we have a sample of students from a statistics class and found that the mean age is 20, then we probably also want to know where the rest of the sample falls. If it ends up that everyone in the sample is 20 years old, then there is no variation, or dispersion, in our sample -- because everyone is the same age, there is no variety in the ages. More likely, though, the mean age will only represent the majority of the class; there will be many others who are <u>not</u> 20 years old. Measures of dispersion indicate how different the rest of the sample is from the average.

- Let's say we have several people in our sample who are not 20 years old, but instead range in age from 17 to 47 years of age. Although we found the mean age to be 20, we also know that there is quite a bit of disparity among the ages. We would expect a high degree of dispersion in the sample with regard to the age variable.
- Like measures of central tendency, there are different measures of dispersion based on the level of measurement of the variable in question. In the example above, we were talking about age, which is an interval-ratio variable. The most appropriate measure of central tendency for an interval-ratio variable is the mean, and the most appropriate measure of dispersion is the standard deviation, but the range (R) and interquartile range (Q) can also be used. For ordinal data we use the median for the measure of central tendency and the range (R) or the interquartile range (Q) to measure dispersion. A simple way to remember these is with the chart below.

| Level of Measurement | Best Measure of Central Tendency | Measure of Dispersion |
| --- | --- | --- |
| Nominal | Mode | None |
| Ordinal | Median | Range (R) & Interquartile Range (Q) |
| Interval-Ratio | Mean (Median for skewed distributions) | Standard Deviation (s), Range (R), & Interquartile Range (Q) |

## 5.2    The Range (R) and Interquartile Range (Q).

- The range measures the distance between the highest and lowest scores in any distribution. It is a "quick and easy" indicator of dispersion in a distribution. However, the range focuses on only the two most extreme values in a distribution.
- We can also find the interquartile range (Q), which focuses on the middle 50% of the cases in the distribution. While the interquartile range corrects partially for the shortcomings of the range (particularly that it is based on only the two most extreme values), it too ignores most of the values in a distribution, focusing still on just two values in the distribution.

## 5.3 Computing the Range and Interquartile Range.

- To calculate the range for numerical variables, simply subtract the lowest category value from the highest – this number is the distance between the highest and lowest scores.
  - The measures of dispersion for ordinal variables are not as simple, primarily because we are trying to squeeze our measure of dispersion to fit this type of data. Ordinal data, remember, are data organized into some hierarchy, from lowest to highest, but exact distances between two scores are unknown. Thus, ordinal data are usually categories such as freshman, sophomore, junior, senior, or scaled items such as strongly agree, agree, disagree, strongly disagree. The appropriate measure of dispersion is the range (R) of scores or categories, such as the scale ranges from strongly agree to strongly disagree, or from freshman to senior.
- To calculate the interquartile range on a numerical variable, subtract the first quartile (or the 25th percentile) from the third quartile (or the 75th percentile): $Q=Q_3-Q_1$. Remember: if our data are in true categories, with no zero point, then the quartiles are the categories, not the frequencies associated with each.
  - An easy way to determine the location of the first and third quartiles is to multiple the sample size by 0.25 for the first quartile and by 0.75 for the third quartile. The resultant value will indicate the cumulative frequency for the appropriate quartile. Then determine which value/category corresponds to that cumulative frequency.

## 5.4 The Standard Deviation and Variance.

- Three characteristics of "good" measures of dispersion include: (1) using all of the scores in a distribution, (2) describing the average distance between scores and the mean, and (3) increasing in value as the variation in the distribution increases.
- The most widely used measure of dispersion is the standard deviation (**s** for samples and **σ** for populations), which can only be used with interval-ratio data because it makes use of the mean.
- Stop for a moment to think about the term "deviants." We all have probably met someone whom we considered a deviant--someone who does not quite fit our definition of "normal." Standard deviations are interested in those people: the ones who are not average -- those who do not fit the norm. The more "abnormal" the sample, the higher the standard deviation will be. Thus, we find the average case and then find out how much our sample deviates from that case.
  - Take a moment to think of all the students in your statistics course. If the average age is 20, how far from the mean are you? How far from the mean is the person next to you? If you subtract your age from the mean age (20), is there a difference or does it equal zero? If that difference is zero, what does that tell you?
- The definitional equation for the standard deviation is:

$$s = \sqrt{\frac{\sum\left(X_i - \overline{X}\right)^2}{N}} \quad or \quad s = \sqrt{\frac{\sum\left(X_i - \overline{X}\right)^2}{N-1}}$$

- First, note that the value $(X_i - \overline{X})$ is called a deviation; every score in a distribution has its own deviation, that is, its own difference from the mean. In this equation each score is subtracted from the mean, squared and summed, and divided by the sample size. However, we use (N-1) in the denominator of the formula when using sample data; technically, N is only used for population data. The square-root of this value indicates the amount of variation there is in the sample. In an age distribution, if everyone is basically the same age, there will be very little variation, but if there are a lot of different ages, there will be a lot of variation. Squaring the standard deviation yields a statistic called the variance ($s^2$ for samples and $\sigma^2$ for populations)

**5.5   Computing the Standard Deviation: An Additional Example**.

**5.6   Interpreting the Standard Deviation.**
- Putting the standard deviation into words will be easier once you understand the concept of the "normal curve" (chapter 5), but at this point you can speak of the dispersion as plus or minus one standard deviation away from the mean: if the mean age is 20, and the standard deviation is 3, we can say that the average age is 20 years, plus or minus three years.
- The standard deviation does not have a set range of possible values. Theoretically, it can be as low as zero (meaning no deviations from the mean--or more simply put: everyone is exactly the same), and as large as infinity. The larger the value, the more dispersed the scores are around the mean. A good rule of thumb to use is that a standard deviation of approximately half the size of the mean represents a reasonable "normal" curve (discussed in greater detail in chapter 5).

**MULTIPLE-CHOICE QUESTIONS**

1.     Deviations are _____.
   a.  the amount of variety in a distribution of scores
   b.  the highest score minus the lowest score
   c.  the distance between the scores and the mean
   d.  changes in the quality of the scores

*Learning Objective: 2*

2.    Dispersion is _____.
      a.  the amount of variety in a distribution of scores
      b.  the highest score minus the lowest score
      c.  the distance between different scores
      d.  changes in the quantity of the scores

      *Learning Objective: 1*


3.    Good measures of dispersion should _____.
      a.  increase in value as the scores become more diverse
      b.  use all of the scores in the distribution
      c.  describe the typical deviation of the scores
      d.  all of these choices

      *Learning Objective: 3*


4.    To calculate deviations _____.
      a.  sum all scores
      b.  subtract the mean from each score
      c.  subtract the mean from each score and then sum
      d.  square the difference of the mean and each score

      *Learning Objective: 2*


5.    As sample size _____, the sum of squared deviations _____.
      a.  increases, decreases
      b.  increases, increases
      c.  decreases, increases
      d.  none of these choices, there is no relationship between the sample size and the sum of squared deviations

      *Learning Objective: 4*


6.    The standard deviation is the best measure of dispersion for _____ level variables.
      a.  all of these choices
      b.  nominal
      c.  ordinal
      d.  interval-ratio

      *Learning Objective: 3*

7. The range is _____.
   a. the highest score minus the lowest score
   b. the distance between the scores and the mean
   c. a measure of dispersion for nominal data
   d. affected by the sample size

   *Learning Objective: 2*


8. Standard deviation(s) is (are) _____.
   a. a measure of dispersion for interval-ratio variables
   b. the square root of the squared deviations of the scores around the mean divided by N
   c. a measure of dispersion for a sample of a population
   d. all of these choices

   *Learning Objective: 2*


9. Variance is _____.
   a. the squared deviations of the scores around the mean divided by N
   b. used primarily for ordinal data
   c. compares the number of cases with the number of categories
   d. all of these choices

   *Learning Objective: 2*


10. Standard deviations range from _____ to _____.
    a. -1 to +1
    b. 0 to 1
    c. 0 to infinity
    d. 0 to 10

    *Learning Objective: 4*


11. A standard deviation of zero indicates _____.
    a. nothing
    b. complete variety in the distribution
    c. the mean and standard deviation are the same
    d. no variety in the distribution

    *Learning Objective: 4*

12. Distributions with much variability or dispersion will be _____.
   a. taller
   b. thinner
   c. flatter
   d. skewed

   *Learning Objective: 1*


13. Which of the following measures of dispersion can be calculated for ordinal level variables?
   a. standard deviation
   b. interquartile range
   c. variance
   d. deviation

   *Learning Objective: 3*


14. If you have an average age of 35 with a standard deviation of 37 years, what might you conclude about the distribution?
   a. The scores are clustered around the mean.
   b. The average age is 35, and the typical difference between each age and the mean is two years.
   c. There is very little variation in the sample on this variable.
   d. There is a lot of variation in the sample on this variable.

   *Learning Objective: 2*


15. If the average score on a statistics exam was 76, and the standard deviation was 7, what can you conclude about the distribution?
   a. The scores are clustered around the mean.
   b. The average score on the exam was 76, and the typical difference between each exam score and the mean is seven points.
   c. There is very little variation in the sample on this variable.
   d. all of these choices

   *Learning Objective: 2*

16.     The _____ focuses only on the middle 50 percent of the cases in a distribution.
        a. range
        b. interquartile range
        c. variance
        d. standard deviation

*Learning Objective: 2*


17.     A "quick and easy" measure of dispersion is _____.
        a. range
        b. interquartile range
        c. variance
        d. standard deviation

*Learning Objective: 2*


18.     The _____ is(are) based on only two scores in a distribution.
        a. range
        b. interquartile range
        c. variance
        d. standard deviation

*Learning Objective: 2*


19.     Measures of dispersion _____ in value as the distribution exhibits _____ variation.
        a. decrease, more
        b. decrease, less
        c. increase, less
        d. decrease, no

*Learning Objective: 1*

20. Use _____ in the denominator of the standard deviation when you have sample data, and use _____ in the denominator of the standard deviation when you have population data.
a. N, N–1
b. N–1, N
c. N, $N^2$
d. $N^2$, N

*Learning Objective: 2*

## WORK PROBLEMS

1. Calculate and interpret the most appropriate measure of dispersion for the *number of children in family* variable.

| Number of Children in Family | Frequency (f) |
|---|---|
| 0 | 20 |
| 1 | 10 |
| 2 | 56 |
| 3 | 32 |
| 4 | 10 |
| 5 | 2 |
| 6 | 0 |
| 7 or more | 0 |
| | N = 130 |

*Learning Objective: 3*

2. Calculate and interpret the most appropriate measure of dispersion for the *social class* variable.

| Social Class | Frequency (f) |
|---|---|
| Lower | 5 |
| Working | 18 |
| Middle | 37 |
| Upper Middle | 55 |
| Upper | 2 |
| | N = 117 |

*Learning Objective: 3*

3. Calculate and interpret the most appropriate measure of dispersion for the *political party affiliation* variable.

| Political Ideology | Frequency (f) |
|---|---|
| Extremely conservative | 10 |
| Conservative | 45 |
| Slightly conservative | 30 |
| Moderate | 50 |
| Slightly liberal | 25 |
| Liberal | 40 |
| Extremely liberal | 20 |
| | N = 220 |

*Learning Objective: 3*

4. Calculate and interpret the most appropriate measure of dispersion for the *class ranking* variable.

| Class Ranking | Frequency (f) |
|---|---|
| Freshmen | 3 |
| Sophomores | 5 |
| Juniors | 15 |
| Seniors | 13 |
| Graduates | 1 |
| | N = 37 |

*Learning Objective: 3*

5. Calculate and interpret the most appropriate measure of dispersion for the *grades on statistics exam* variable.

| Grades on Statistics Exam | Frequency (f) |
|---|---|
| 100 | 2 |
| 95 | 12 |
| 85 | 15 |
| 75 | 18 |
| 65 | 15 |
| 55 | 5 |
| | N = 67 |

*Learning Objective: 3*

6. Calculate and interpret the most appropriate measure of dispersion for the *incomes of recent college graduates* variable.

**Incomes of Recent Graduates of A-1 College**

| | | | | |
|---|---|---|---|---|
| $12,500 | $13,500 | $17,500 | $30,000 | $23,000 |
| $14,350 | $18,900 | $32,450 | $14,000 | $27,650 |
| $33,450 | $26,760 | $19,000 | $21,000 | $37,800 |
| $12,000 | $22,500 | $27,500 | $23,000 | $20,900 |

*Learning Objective: 3*

7. Using the *incomes of recent college graduates* data from problem 6, calculate the range.

*Learning Objective: 2*

8. Using the same information on recent college graduates' incomes, calculate the interquartile range.

*Learning Objective: 2*

9. A set of twenty new hires at a marketing firm all receive the same starting salary. What is the standard deviation for their salaries?

*Learning Objective: 2*

10. Calculate and interpret the standard deviation for the *age* variable.

| Age | Frequency (f) |
|---|---|
| 18 | 3 |
| 19 | 5 |
| 20 | 8 |
| 21 | 3 |
| 37 | 1 |
| | N = 20 |

*Learning Objective: 3*

11. Using the age data from problem 10, calculate and interpret the range.

*Learning Objective: 2*

12. Using the age data from problem 10, calculate and interpret the interquartile range.

*Learning Objective: 2*

13. Compare the three dispersion statistics you calculated in problems 10 through 12. Which do you think is the best indicator of dispersion, and why?

*Learning Objective: 2 and 3*

**SPSS or MicroCase WORK PROBLEMS**

The following work problems can be completed using either SPSS or MicroCase. The solutions will be the same regardless of which one you use. Your instructor may have a preference, so be sure to check with him/her prior to beginning these problems.

1. Find the appropriate measure of dispersion for the variables *age* and *tvhours*. Write up a brief summary of each one.

2. Find the appropriate measure of dispersion for the variables *marital* and *partyid*. Write up a brief summary of each one.

3. Find the appropriate measure of dispersion for the variables *attend* and *polviews*. Write up a brief summary of each one.

4. Find the appropriate measure of dispersion for *realinc* and *sei*. Write a summary of the comparison of the two distributions. Which variable shows more dispersion?

## ANSWER KEY

### MULTIPLE-CHOICE QUESTIONS

| | | | | | | |
|---|---|---|---|---|---|---|
| 1. | c | (p. 109) | 11. | d | (p. 115) |
| 2. | a | (p. 105) | 12. | c | (p. 106) |
| 3. | d | (p. 109) | 13. | b | (p. 118) |
| 4. | b | (p. 109) | 14. | d | (p. 115) |
| 5. | b | (p. 110) | 15. | d | (p. 115) |
| 6. | d | (p. 119) | 16. | b | (p. 107) |
| 7. | a | (p. 106) | 17. | a | (p. 106) |
| 8. | d | (p. 108-111) | 18. | d | (p. 106-108) |
| 9. | a | (p. 110) | 19. | b | (p. 106) |
| 10. | c | (p. 115) | 20. | b | (p. 110) |

### WORK PROBLEMS

1.   Because the data are interval-ratio, the most appropriate measure of dispersion is the standard deviation.

| Number of Children ($X_i$) | $(X_i - \overline{X})^2$ | | | | $\times f$ | | |
|---|---|---|---|---|---|---|---|
| 0 | $(0\text{-}2.06)^2 = (-2.06)^2 =$ | 4.24 | | $\times 20$ | $=$ | 84.87 |
| 1 | $(1\text{-}2.06)^2 = (-1.06)^2 =$ | 1.12 | | $\times 10$ | $=$ | 11.24 |
| 2 | $(2\text{-}2.06)^2 = (-0.06)^2 =$ | 0.004 | | $\times 56$ | $=$ | 0.20 |
| 3 | $(3\text{-}2.06)^2 = (0.94)^2 =$ | 0.88 | | $\times 32$ | $=$ | 28.28 |
| 4 | $(4\text{-}2.06)^2 = (1.94)^2 =$ | 3.76 | | $\times 10$ | $=$ | 37.64 |
| 5 | $(5\text{-}2.06)^2 = (2.94)^2 =$ | 8.64 | | $\times 2$ | | 17.29 |
| 6 | $(6\text{-}2.06)^2 = (3.94)^2 =$ | 15.52 | | $\times 0$ | $=$ | 0 |
| 7 or more | $(7\text{-}2.06)^2 = (4.94)^2 =$ | 24.40 | | $\times 0$ | $=$ | 0 |
| | | | | $\Sigma = 130$ | | $\Sigma = 179.51$ |

$$\overline{X} = \frac{\sum X_i}{N} = \frac{(0\times20)+(1\times10)+(2\times56)+(3\times32)+(4\times10)+(5\times2)+(6\times0)+(7\times0)}{130}$$

$$\overline{X} = \frac{268}{130} = 2.06$$

$$s = \sqrt{\frac{\sum(X_i - \overline{X})^2}{N-1}} = \sqrt{\frac{179.51}{130-1}} = \sqrt{1.39} = 1.18$$

← Sample

Therefore, the sample has a mean of 2.06 children, and a standard deviation of 1.18 children.

99

2. This variable is ordinal, and as such, it is a sticky variable to find the measure of dispersion. It is possible to talk about the range of the values; the range is between upper class and lower class. In addition to the range, the interquartile range is appropriate for ordinal-level variables. The first quartile occurs at the 29th case (25%×117=29.25), and the third quartile occurs at the 88th case (75%×117=87.75). Therefore, $Q_1$ occurs in the "middle" category, and $Q_3$ occurs in the "upper middle" category. The interquartile range is the difference between these two scores.

3. This variable is ordinal too. Two measures of dispersion are appropriate for ordinal data, the range and interquartile range. The range is simply the difference between the highest and lowest values; therefore, the range in political ideology is between extremely conservative and extremely liberal ideologies. In order to calculate the interquartile range we must identify the first and third quartiles. $Q_1$ occurs at the 55th case (25%×220=55), and $Q_3$ occurs at the 165th case (75%×220=165). The 55th case is in the "conservative" category, and the 165th case is in the "liberal" category. The interquartile range is the distance between the conservative and liberal ideologies.

4. Class ranking is ordinal level data. The range is the difference between the lowest category, "freshmen," and the highest category, "graduates." The interquartile range is the difference between the first quartile ($Q_1 = 25\% \times 37 = 9.25$ or 9th case) and the third quartile ($Q_3 = 75\% \times 37 = 27.75$ or 28th case). The 9th case is a "junior," and the 28th case is a "senior;" therefore, the interquartile range is the difference between "juniors" and "seniors."

5. This variable is interval-ratio, so we can compute the standard deviation.

| Grades ($X_i$) | $\left(X_i - \overline{X}\right)^2$ | | | $\times f$ | | |
|---|---|---|---|---|---|---|
| 100 | $(100\text{-}77.8)^2 =$ | $(22.2)^2$ | $= 492.84$ | $\times 2$ | $=$ | 985.68 |
| 95 | $(95\text{-}77.8)^2 =$ | $(17.2)^2$ | $= 295.84$ | $\times 12$ | $=$ | 3550.08 |
| 85 | $(85\text{-}77.8)^2 =$ | $(7.2)^2$ | $= 51.84$ | $\times 15$ | $=$ | 777.60 |
| 75 | $(75\text{-}77.8)^2 =$ | $(\text{-}2.8)^2$ | $= 7.84$ | $\times 18$ | $=$ | 141.12 |
| 65 | $(65\text{-}77.8)^2 =$ | $(\text{-}12.8)^2$ | $= 163.84$ | $\times 15$ | $=$ | 2457.60 |
| 55 | $(55\text{-}77.8)^2 =$ | $(\text{-}22.8)^2$ | $= 519.84$ | $\times 5$ | $=$ | 2599.20 |
| | | | | $\Sigma = 67$ | | $\Sigma = 10511.28$ |

$$\overline{X} = \frac{\sum X_i}{N} = \frac{(100 \times 2) + (95 \times 12) + (85 \times 15) + (75 \times 18) + (65 \times 15) + (55 \times 5)}{67}$$

$$\overline{X} = \frac{5215}{67} = 77.8$$

$$s = \sqrt{\frac{\sum(X_i - \overline{X})^2}{N-1}} = \sqrt{\frac{10511.28}{67-1}} = \sqrt{159.26} = 12.6$$

Therefore, the mean grade on the statistics exam was 77.8, with a standard deviation of 12.6 points.

6.   Here again we have interval-ratio data, so we should compute the standard deviation. If we square the difference between the scores and the mean, we are going to get a very large number. Thus, here's a quick trick: divide each score by 1,000 and work with these numbers. But don't forget to multiply by 1,000 at the end of the problem.

$$\overline{X} = \frac{\sum X_i}{N} = \frac{447.76}{20} = 22.388 \quad or \quad \$22,388$$

$$s = \sqrt{\frac{\sum(X_i - \overline{X})^2}{N-1}} = \sqrt{\frac{1064.11}{20-1}} = \sqrt{56.01} = 7.484 \quad or \quad \$7,484$$

| Income of Recent Graduates ($X_i$) | $(X_i - \overline{X})^2$ | | |
|---|---|---|---|
| 12 | $(12\text{-}22.388)^2$ | $= (-10.39)^2$ | $= 107.91$ |
| 12.5 | $(12.5\text{-}22.388)^2$ | $= (-9.89)^2$ | $= 97.77$ |
| 13.5 | $(13.5\text{-}22.388)^2$ | $= (-8.89)^2$ | $= 79.00$ |
| 14 | $(14\text{-}22.388)^2$ | $= (-8.39)^2$ | $= 70.36$ |
| 14.35 | $(14.35\text{-}22.388)^2$ | $= (-8.04)^2$ | $= 64.61$ |
| 17.5 | $(17.5\text{-}22.388)^2$ | $= (-4.89)^2$ | $= 23.89$ |
| 18.9 | $(18.9\text{-}22.388)^2$ | $= (-3.49)^2$ | $= 12.17$ |
| 19 | $(19\text{-}22.388)^2$ | $= (-3.39)^2$ | $= 11.48$ |
| 20.9 | $(20.9\text{-}22.388)^2$ | $= (-1.49)^2$ | $= 2.21$ |
| 21 | $(21\text{-}22.388)^2$ | $= (-1.39)^2$ | $= 1.93$ |
| 22.5 | $(22.5\text{-}22.388)^2$ | $= (0.11)^2$ | $= 0.01$ |
| 23 | $(23\text{-}22.388)^2$ | $= (0.61)^2$ | $= 0.37$ |
| 23 | $(23\text{-}22.388)^2$ | $= (0.61)^2$ | $= 0.37$ |
| 26.76 | $(26.76\text{-}22.388)^2$ | $= (4.37)^2$ | $= 19.11$ |
| 27.5 | $(27.5\text{-}22.388)^2$ | $= (5.11)^2$ | $= 26.13$ |
| 27.65 | $(27.65\text{-}22.388)^2$ | $= (5.26)^2$ | $= 27.69$ |
| 30 | $(30\text{-}22.388)^2$ | $= (7.61)^2$ | $= 57.94$ |
| 32.45 | $(32.45\text{-}22.388)^2$ | $= (10.06)^2$ | $= 101.24$ |
| 33.45 | $(33.45\text{-}22.388)^2$ | $= (11.06)^2$ | $= 122.37$ |
| 37.8 | $(37.8\text{-}22.388)^2$ | $= (15.41)^2$ | $= 237.53$ |
| | | $\Sigma =$ | $1064.11$ |

Therefore, the mean income of recent college graduates is $22,388, with a standard deviation of $7,484.

7.    The range is the difference of the highest and lowest scores. The highest income was $37,800, and the lowest income was $12,000. Therefore, the range equals $25,800 (37800-12000=25800). The difference between the highest and lowest incomes was $25,800.

8.    The first quartile is located at the $5^{th}$ case (25%×20=5), which is an income of $14,350. The third quartile is located at the $15^{th}$ case (75%×20=15), which is an income of $27,500. Because income is a "numerical" variable, we can subtract these two values. Therefore, the interquartile range, or the distance between the first and third quartiles, is $13,150 (27500-14350=13150).

9.    Because all of the salaries are the same, the standard deviation is zero.

10.   Again, we need to compute the standard deviation for interval-ratio data.

| Age ($X_i$) | $\left(X_i - \overline{X}\right)^2$ | | × f | |
|---|---|---|---|---|
| 18 | $(18-20.45)^2 = (-2.45)^2 =$ | 6.00 | × 3 = | 18.01 |
| 19 | $(19-20.45)^2 = (-1.45)^2 =$ | 2.10 | × 5 = | 10.51 |
| 20 | $(20-20.45)^2 = (-0.45)^2 =$ | 0.20 | × 8 = | 1.62 |
| 21 | $(21-20.45)^2 = (0.55)^2 =$ | 0.30 | × 3 = | 0.91 |
| 37 | $(37-20.45)^2 = (16.55)^2 =$ | 273.90 | × 1 = | 273.90 |
| | | | $\Sigma = 20$ | $\Sigma = 304.95$ |

$$\overline{X} = \frac{\sum X_i}{N} = \frac{(18 \times 3) + (19 \times 5) + (20 \times 8) + (21 \times 3) + (37 \times 1)}{20} = \frac{409}{20} = 20.45$$

$$s = \sqrt{\frac{\sum (X_i - \overline{X})^2}{N-1}} = \sqrt{\frac{304.95}{20-1}} = \sqrt{16.05} = 4.0$$

Therefore, the mean age is 20.45 years, with a standard deviation of 4 years.

11.   The oldest person is 37, and the youngest is 18. Therefore, the range, or the distance between the oldest and youngest people in the sample, is 19 (37-18=19).

12.   The first quartile is located at the $5^{th}$ case (25%×20=5), which is an age of 19. The third quartile is located at the $15^{th}$ case (75%×20=15), which is an age of 20. Therefore, the interquartile range, or the distance between the first and third quartiles, is 1 (20-19=1).

13.  In problem 10, the standard deviation equaled 4. In problem 11, the range equaled 19. In problem 12, the interquartile range equaled 1. As expected, the range has been influenced by the single extreme age value of 37. The interquartile range is therefore a slightly better indicator of the range of ages in the distribution. But in the end, the standard deviation is the best indicator of dispersion, with its focus on deviation from the mean.

## SPSS or MicroCase WORK PROBLEMS

If you had problems with any of these questions, be sure to re-read the appropriate sections at the end of chapter 4 in your textbook.

1.  The two variables (*tvhours* and *age*) are interval ratio level variables, and as such, all three measures of dispersion are appropriate, but the standard deviation is the best measure. In order to save space, I'll use the "Statistics" option of the "Frequencies" command to obtain just the means and standard deviations of these variables. I request the mean (which is a measure of central tendency, not a measure of dispersion) because it assists in the interpretation of the standard deviation. This output appears below.

**Statistics**

|  |  | AGE: AGE OF RESPONDENT | TVHOURS: HOURS PER DAY WATCHING TV |
|---|---|---|---|
| N | Valid | 1493 | 673 |
|  | Missing | 7 | 827 |
| Mean |  | 47.43 | 2.96 |
| Std. Deviation |  | 16.912 | 2.346 |
| Variance |  | 286.027 | 5.506 |

The mean age of the respondents is 47.43 years, with a standard deviation of 16.9 years.

The mean hours of television watch a day for this sample is 2.96. The standard deviation is 2.346 hours. There appears to be a lot of variation in the hours of TV watched per day, yielding a fairly high standard deviation in comparison to the mean.

2.  Because *marital* and *partyid* are nominal variables, there are no appropriate measures of dispersion available.

3.    *Attend* and *polviews* are ordinal variables.  Their frequency distributions appear below.

**ATTEND:  HOW OFTEN R ATTENDS RELIGIOUS SERVICES**

| | | Frequency | Percent | Valid Percent | Cumulative Percent |
|---|---|---|---|---|---|
| Valid | LESS THAN 1 A YEAR | 441 | 29.4 | 29.5 | 29.5 |
| | 1 A YR-SEV TIMES WK | 347 | 23.1 | 23.2 | 52.7 |
| | 1 A MNTH-ALMST WKLY | 322 | 21.5 | 21.6 | 74.3 |
| | EVERY WEEK + | 384 | 25.6 | 25.7 | 100.0 |
| | Total | 1494 | 99.6 | 100.0 | |
| Missing | 9 | 6 | .4 | | |
| Total | | 1500 | 100.0 | | |

**POLVIEWS:  THINK OF SELF AS LIBERAL OR CONSERVATIVE**

| | | Frequency | Percent | Valid Percent | Cumulative Percent |
|---|---|---|---|---|---|
| Valid | LIBERAL | 384 | 25.6 | 26.7 | 26.7 |
| | MODERATE | 571 | 38.1 | 39.7 | 66.4 |
| | CONSERVATIVE | 483 | 32.2 | 33.6 | 100.0 |
| | Total | 1438 | 95.9 | 100.0 | |
| Missing | 9 | 62 | 4.1 | | |
| Total | | 1500 | 100.0 | | |

The range and interquartile range are appropriate measures of dispersion for ordinal variables.  The range for *attend* is the difference between the highest and lowest categories, or the difference between "every week +" and "less than 1 a year."
In order to determine the interquartile range, we must first locate the first and third quartiles.  The first quartile (a cumulative percent of 25) occurs somewhere in the "less than 1 a year" category, and the third quartile (a cumulative percent of 75) occurs somewhere in the "every week +" category.  The interquartile range is the difference between the "every week +" and "less than 1 a year" categories.

The range for *polviews* is the difference between the highest and lowest categories, or the difference between "conservative" and "liberal."  In order to determine the interquartile range, we must first locate the first and third quartiles.  The first quartile (a cumulative percent of 25) occurs somewhere in the "liberal" category, and the third quartile (a cumulative percent of 75) occurs somewhere in the "conservative" category.  The interquartile range is the difference between the "conservative" and "liberal" categories.

4. Both *sei* and *realinc* are interval-ratio level variables; therefore the standard deviation is the best measure of dispersion. (And as I did above, I request the mean in order to make the interpretation of the standard deviation clearer.) The output appears below.

**Statistics**

| | | REALINC: FAMILY INCOME IN CONSTANT $ | SEI: RESPONDENT SOCIOECONOMIC INDEX |
|---|---|---|---|
| N | Valid | 1282 | 1426 |
| | Missing | 218 | 74 |
| Mean | | 33049.90 | 50.017 |
| Std. Deviation | | 31612.529 | 19.7090 |

The mean family income is $33,049.90, with a standard deviation of $31,612.529, and the mean value on the socioeconomic index is 50.017 points, with a standard deviation of 19.709 points. The standard deviation for income is almost as large as the mean itself, where the standard deviation for SEI is less than half the mean. Therefore, income shows much more variation than the socioeconomic index.

# CHAPTER SIX
## The Normal Curve

_Learning Objectives_: _By the end of this chapter, you will be able to:_

1.  _Define and explain the concept of the normal curve._
2.  _Convert empirical scores to Z scores and use Z scores and the normal curve table (Appendix A) to find areas above, below, and between points on the curve._
3.  _Express areas under the curve in terms of probabilities._


## KEY TERMS

- ✓ Normal curve  (p. 127)
- ✓ Normal curve table  (p. 131)
- ✓ Probability  (p. 137)
- ✓ Z scores  (p. 130)


## CHAPTER SUMMARY

In chapter three you learned about histograms, which are used to graphically display the frequencies of interval-ratio variables.  In histograms, the frequencies are displayed with contiguous bars, each one marking the frequency of each category of the variable in question.  Now, consider drawing a curve that copies the shape of that distribution.  If that curve is bell-shaped, with one mode (unimodal), and the median and mean are about the same place as the mode, we could conclude that this variable has a normal distribution and is shaped by a normal curve.

Theoretically, a normal curve should be perfectly symmetrical, with 50% of the cases falling on each side of the highest point, which _theoretically_ is the mean, median, and mode.  It would be very rare indeed for a distribution of any variable to exactly model the theoretical normal curve, but it is important for the assumption of normality.  With this assumption met, we can describe our distribution, find the areas above and below the mean, as well as between scores, and make predictions.

# CHAPTER OUTLINE

## 6.1 Introduction.

- When describing a distribution, we previously used the mean and standard deviation. With the assumption of the normal curve (normality) met, we can now describe our distribution in terms of the percentage of respondents above and below the mean (or any other point in the distribution). In the case of a normal curve, we can divide the area under the curve into segments, each segment representing one standard deviation. Therefore, we can find the area under the normal curve, which corresponds to the area of one standard deviation away from the mean (plus or minus). Luckily for us, this area has already been found: 68.26% of the respondents will <u>always</u> fall between plus and minus one standard deviation from the mean. This means that 68.26% of the respondents are within the area of one standard deviation below the mean <u>and</u> one standard deviation above the mean.

- We are also given the areas that are two and three standard deviations away from the mean. The area under the normal curve two standard deviations away from the mean encompasses 95.44% of the sample; the area under the normal curve three standard deviations away from the mean encompasses 99.72% of the sample. These will always be the same for any normal curve. (Therefore I would suggest memorizing them.)

- If 68.26% of the sample falls within one standard deviation of the mean, then we also know that the majority of the sample is in that area. Thus, the majority of the respondents cluster around the mean, giving the normal curve its shape. Also, if 99.72% of all the scores lie within three standard deviations from the mean, then only a very small percentage (0.28%) fall beyond three standard deviations from the mean. Thus, very extreme scores are very rare, also giving the normal curve its shape (in particular, the low "tails" at the lower and upper ends of the normal curve).

## 6.2 Computing Z Scores.

- If we know that the mean age in a statistics course is 20 years old, with a standard deviation of three years, and you are 23 years old, we can determine the percentage of the rest of the class that is younger than you, the percentage of the class that is older than you, and how far above the average age you are. We do all of this by making use of Z scores.

- Z scores are *standardized scores* in a distribution, enabling us to make comparisons between variables measured in different scales. Basically it works like this: in the U.S. we use the Fahrenheit scale to measure temperature, and it has a freezing point of $32^\circ$. In other parts of the world the Celsius scale is used which has a freezing point of zero. If we were to change our description of the temperature outside today from Fahrenheit to Celsius, we are not changing the temperature, but only the **scale** on which we are measuring it. Z scores work the same way: we are not changing the scores as much as changing the scale in which we are measuring the variable.

- Z scores will always have a mean of zero and a standard deviation of one. Thus in the example from the statistics class, if the raw scores have a mean age of 20 years old and a standard deviation of 2.5 years, the Z scores of the age variable have a mean age of zero and a standard deviation of one.

- To convert any score into a Z score, we use the following equation:

$$Z = \frac{X_i - \overline{X}}{s}$$

- This equation converts any score into a Z score. Thus if you wanted to convert your own age into a Z score, you would plug your age into $X_i$ and complete the equation. If you were 23 years old, the equation would be:

$$Z = \frac{X_i - \overline{X}}{s} = \frac{23 - 20}{2.5} = \frac{3}{2.5} = +1.2$$

- Therefore, if you are 23 years old, in a class with the mean age of 20 years old and a standard deviation of 2.5 years, you are 1.2 standard deviation units <u>above</u> the mean.

### 6.3 The Normal Curve Table.
- Immediately above, we found that someone 23 years old in a class with a mean age of 20 and a standard deviation of 2.5, was 1.2 standard deviations above the mean. Well, so what, you might say. So what, indeed! Now the fun begins. If you look at Appendix A in the back of your textbook, you will see several columns. These columns identify the area under the normal curve. Find the Z score of 1.20 in column a. In column b, we find that the area between the mean and Z is 0.3849, or 38.49% of the respondents fall between the mean (20 years old) and your Z score of 1.20 (23 years of age). To find the area above your score (or those who are older than 23), go to column c; 0.1151 or 11.51% of the class is older than 23.

### 6.4 Finding Total Area Above and Below a Score.
- Z scores can also be used to find the total proportion or percentage of scores which fall below a certain score. Remember that 50% of all scores fall above and below the mean in the theoretical normal curve. As such, we know that 50% of the cases will always be below the mean and 50% will always be above the mean. If we have a positive Z score, then we can find the area between the mean and Z, and add 50% to find the total area below that score. Thus, if you are 23 years old in the above example, 38.49% of the sample falls between your age and the mean, and 88.49% of the sample is younger than you (50 + 33.49 = 88.49%).

### 6.5 Finding Areas Between Two Scores.
- If you want to find the area <u>between</u> two scores, let's say your age of 23 and someone who is 18 years old, you need to convert both scores to Z scores and add the corresponding areas. We know the area between the mean and 23 is 38.49%. First, find the Z score for 18:

$$Z = \frac{X_i - \overline{X}}{s} = \frac{18 - 20}{2.5} = \frac{-2}{2.5} = -0.80$$

- Someone who is 18 years old in this class is 0.8 standard deviation units below the mean. (We say that the person lies below the mean because of the negative value for the Z score.) The area between the mean and the age of 18 is 0.2881 (column b), or 28.81% of the sample fall between the ages of 18 and 20.
- We can now find the area between these two scores by adding the areas between each score and the mean:

$$28.81\% + 38.49\% = 67.3\%$$

- Therefore, 67.3% of the sample falls between the ages of 18 and 23.

## 6.6 Using the Normal Curve to Estimate Probabilities.

- We can also make predictions with Z scores. Because the area between the mean and Z is in proportions, we can consider, for example, the likelihood that a randomly selected respondent will be between the age of 20 and 22. To do so, we begin by finding the Z score:

$$Z = \frac{X_i - \overline{X}}{s} = \frac{22 - 20}{2.5} = \frac{2}{2.5} = +0.80$$

- The corresponding area for a Z score of 0.80 (column b) is 0.2881. Proportions range from 0 to 1 (with zero meaning no chance at all and 1 meaning a certainty), therefore, the likelihood that a randomly selected student from our sample will be between the ages of 20 and 22 is 0.2282.

## MULTIPLE-CHOICE QUESTIONS

1. Z scores are expressed in _____ units.
   a. standard error
   b. mean
   c. standard deviation
   d. raw

   *Learning Objective: 2*

2. The normal curve is _____.
   a. a theoretical distribution of scores
   b. unimodal
   c. bell-shaped
   d. all of these choices

   *Learning Objective: 1*

3.   A normal curve will have a mean of _____ and a standard deviation of _____.
     a. 0, 0
     b. 0, 1
     c. 1, 0
     d. 1, 1

     *Learning Objective: 1*


4.   Z scores are _____.
     a. unidimensional scores
     b. nondenominational scores
     c. rounded scores
     d. standardized scores

     *Learning Objective: 2*


5.   In the Z equation, the mean is subtracted from _____.
     a. an unknown score
     b. a given score
     c. itself
     d. all other scores

     *Learning Objective: 2*


6.   In the Z equation, the denominator is _____.
     a. the standard deviation
     b. one
     c. both a and b
     d. none of the above

     *Learning Objective: 2*


7.   To find the area between a score and the mean, we use the _____ column in Appendix A.
     a. (a)
     b. (b)
     c. (c)
     d. (d)

     *Learning Objective: 2*

8. To find the area below a positive Z score, you need to _____.
   a. subtract the mean from the Z score
   b. find the corresponding Z
   c. add 50% to the corresponding Z
   d. subtract 50% from the corresponding Z

   *Learning Objective: 2*


9. To find the area below a negative Z score, you need to _____.
   a. find the corresponding Z
   b. find the area beyond the Z
   c. find the area between the mean and Z
   d. add 50% to the corresponding Z

   *Learning Objective: 2*


10. To find the area above a negative Z score, you need to _____.
    a. subtract the mean from the Z score
    b. find the corresponding Z
    c. add 50% to the corresponding Z
    d. subtract 50% from the corresponding Z

    *Learning Objective: 2*


11. To find the area between two scores on different sides of the mean, you need to _____.
    a. find the corresponding Z scores
    b. find the area between each Z score and the mean
    c. add the two corresponding areas
    d. all of these choices

    *Learning Objective: 2*


12. Using Appendix A in your book, what is the area between the mean and a Z of 1.71?
    a. 50%
    b. 4.36%
    c. 1.71%
    d. 45.64%

    *Learning Objective: 2*

13. What percent of the sample lies within one standard deviation from the mean?
   a. 50%
   b. 68.26%
   c. 95.44%
   d. 99.72%

   *Learning Objective: 2*


14. What percent of the sample lies within two standard deviations from the mean?
   a. 50%
   b. 68.26%
   c. 95.44%
   d. 99.72%

   *Learning Objective: 2*


15. Using Appendix A in your book, what is the total area above the mean?
   a. 50%
   b. 1%
   c. 0.01%
   d. 45.62%

   *Learning Objective: 2*


16. What is the total area under the normal curve?
   a. 0%
   b. 1%
   c. 50%
   d. 100%

   *Learning Objective: 1*


17. In a normal distribution of 1,000 cases, _____ cases will be between the mean and one standard deviation above the mean?
   a. 34.13
   b. 341.3
   c. 682.6
   d. 500

   *Learning Objective: 2*

18. In a normal distribution of exam grades, where the mean grade was 75 and the standard deviation was 5, an exam grade of 82 has a Z score of _____.
   a. +0.71
   b. +7.00
   c. +1.40
   d. -1.40

   *Learning Objective: 2*


19. Areas between means and positive Z scores and areas between means and negative Z scores are _____.
   a. the opposite of each other
   b. always sum to 100%
   c. always sum to 50%
   d. the same as each other

   *Learning Objective: 2*


20. _____ Z scores are found on the _____ of the normal curve.
   a. negative, right
   b. negative, left
   c. positive, left
   d. all, right

   *Learning Objective:*


## WORK PROBLEMS

(Assume that all distributions are normal, unless stated otherwise.)

1. A researcher has data on starting salaries of recent social science graduates. The mean salary is $27,654, with a standard deviation of $4,569. Assuming a normal distribution, find the corresponding Z scores for the following salaries.
   a. $35,456
   b. $23,875
   c. $27,450
   d. $25,000
   e. $42,750

   *Learning Objective: 2*

2.	Using the same data from problem 1, what percent of the cases is above each of the Z scores you already calculated?

*Learning Objective: 2*


3.	Using the same data from problem 1, what percent of the cases is below each of the Z scores you already calculated?

*Learning Objective: 2*


4.	If there are 10,000 cases in problem 1's distribution, how many cases are above and how many are below each Z score?

*Learning Objective: 2*


5.	Using the same data from problem 1, what percentage of the sample falls below the salary of $30,000?

*Learning Objective: 2*


6.	Bill and Jeff are graduating in a few months and both have found employment. Bill found a job that he loves, but the pay is below average ($25,000); Jeff isn't as happy with his new job, but because the pay is high, he decided to take it ($33,500). What percentage of their class will be receiving salaries between theirs? Use the data from Question 1 above.

*Learning Objective: 2*


7.	The mean years of education in a sample of business executives is 17.2 years, with a standard deviation of 2.3 years. A person graduating with a bachelor's (16 years of education) is hoping to be a business executive someday. What percent of the cases is between the mean and 16 years of education?

*Learning Objective: 2*

8.  Sue is advanced for her age and graduated high school at the age of 12. She is now a sophomore in college at the ripe old age of 14, but realizes she is quite a bit younger than her cohort, whose mean age is 19.8, with a standard deviation of 1.3. How much younger than the rest of the class is she? What percentage of her class is younger than she?

    *Learning Objective: 2*

9.  Poor Sue, being the bright young woman she is, she realizes that the majority of the class is older than she is. What percentage of the class is between 18.5 years and 21.1 years of age?

    *Learning Objective: 2*

10. If the average grade on the first statistics exam is 76 with a standard deviation of 5.9, what percentage of students scored received C's and B's (that is, what percentage of students scored between 70 and 89)?

    *Learning Objective: 2*

11. Sue, whom we met above, earned a 92 on the first exam. How far above the mean is her score? What percentage of the class had a higher score? What percentage of the class had a lower score?

    *Learning Objective: 2*

12. Jeff, who teased Sue about being so young, earned a 67 on the first exam. How far from the mean was his score? What percentage of scores fell between Jeff and Sue's scores?

    *Learning Objective: 2*

13. The average number of children for a sample of Americans is 2.3, with a standard deviation of 1.7. What percent of the sample have had more than three children?

    *Learning Objective: 2*

14. What is the probability that a randomly selected person from our sample will have between 1.9 and 2.3 children?

    *Learning Objective: 3*

15. The employees at Davis and Davis live an average of 10.3 miles from the office (s = 5.3 miles). The CEO is considering moving the office to a new location further from the city center. The new office space will still be an average of 10.3 mile commute for the employees (s=5.3), but who will benefit the most? How far away from the average commute is each person's commute for both the former and new offices?

|  | Former Office | New Office |
|---|---|---|
| Donna | 12.3 | 9.5 |
| Jim | 10.8 | 14.8 |
| Judith | 3.0 | 6.4 |
| Jose | 14.5 | 11.5 |
| Robert | 2.5 | 4.7 |

*Learning Objective: 2*

16. The average SAT score for entering freshmen at a large university is 1100, with a standard deviation of 250. Bill knows he needs to be within one standard deviation of the mean to get accepted. What is the lowest score he can earn on the SATs?

*Learning Objective: 2*

17. Ruth recently earned a 1230 on the SATs and wants to go to the same school as Bill. She knows she's above the minimum requirement, but she wants to know what percent of the entering freshmen fell below her score and what percent fell above. Help her out.

*Learning Objective: 2*

18. Carl has eyes for Ruth. He knows she is very bright and wants to make a good impression on her. He recently found out that her score was 1230 on the SATs, while his were only 1150. What percentage of the other applicants are between their scores?

*Learning Objective: 2*

19. Using the SAT data from the previous problems, what's the probability of someone scoring above Ruth and Carl?

*Learning Objective: 3*

## SPSS or MicroCase WORK PROBLEMS

The following work problems can be completed using either SPSS or MicroCase. The solutions will be the same regardless of which one you use. Your instructor may have a preference, so be sure to check with him/her prior to beginning these problems.

1. Using SPSS or MicroCase find the Z scores for the variable *tvhours*. [Select the "Analyze" menu, select the "Descriptive Statistics" option, and then select "Descriptives." In the Descriptives window, select the *tvhours* variable and then click on the "Save standardized values as variables" box.] Output the mean and standard deviation for both *tvhours* and *ztvhours*.

2. Looking at the dataset, find the raw scores for both *tvhours* and *ztvhours* . Using only the first 25 cases for each variable discuss the implications of the z scores for each one. (Many of the first 20 cases have missing data on these two variables. Just ignore the cases with missing data.)

3. Using the z scores given to you by the computer and Appendix A in the back of your textbook, what percent of the sample falls between each of the first 20 z scores and the mean. (Many of the first 15 cases have missing data on *ztvhours*. Just ignore those cases.)

4. Using the z scores given to you by the computer and Appendix A in the back of your textbook, what percent of the sample falls above each of the first 20 z scores. (Many of the first 15 cases have missing data on *ztvhours*. Just ignore those cases.)

5. Using the z scores given to you by the computer and Appendix A in the back of your textbook, what percent of the sample below above each of the first 20 z scores. (Many of the first 15 cases have missing data on *ztvhours*. Just ignore those cases.)


## ANSWER KEY


## MULTIPLE-CHOICE QUESTIONS

| | | | | | | |
|---|---|---|---|---|---|---|
| 1. | c | (p. 130) | 11. | d | (p. 136) |
| 2. | b | (p. 127) | 12. | d | (p. Appendix A) |
| 3. | d | (p. 130) | 13. | b | (p. 129) |
| 4. | b | (p. 130) | 14. | c | (p. 129) |
| 5. | b | (p. 130) | 15. | a | (p. Appendix A) |
| 6. | a | (p. 130) | 16. | d | (p. 129) |
| 7. | b | (p. 131-134) | 17. | b | (p. 129) |
| 8. | c | (p. 134) | 18. | c | (p. 130) |
| 9. | b | (p. 134) | 19. | d | (p. 131-132) |
| 10. | c | (p. 134) | 20. | b | (p. 128) |

## WORK PROBLEMS

1.   a.  $Z = \dfrac{X_i - \overline{X}}{s} = \dfrac{35456 - 27654}{4569} = +1.71$, or 1.71 standard deviations above the mean

      b.  $Z = \dfrac{X_i - \overline{X}}{s} = \dfrac{23875 - 27654}{4569} = -0.83$, or 0.83 standard deviations below the mean

      c.  $Z = \dfrac{X_i - \overline{X}}{s} = \dfrac{27450 - 27654}{4569} = -0.045$, or 0.045 standard deviations below the mean

      d.  $Z = \dfrac{X_i - \overline{X}}{s} = \dfrac{25000 - 27654}{4569} = -0.58$, or 0.58 standard deviations above the mean

      e.  $Z = \dfrac{X_i - \overline{X}}{s} = \dfrac{42750 - 27654}{4569} = +3.30$, or 3.30 standard deviations above the mean

2.   a.   The area above Z = +1.71 is 4.36%.
      b.   The area above Z = -0.83 is 79.67% (29.67% + 50% = 79.67%).
      c.   The area above Z = -0.045 is 51.99% (1.99% + 50% = 51.99%).
      d.   The area above Z = -0.58 is 71.90% (21.90% + 50% = 71.90%).
      e.   The area above Z = +3.30 is 0.05%.

3.   a.   The area below Z = +1.71 is 95.64% (45.64% + 50% = 95.64%).
      b.   The area below Z = -0.83 is 20.33%.
      c.   The area below Z = -0.045 is 48.01%.
      d.   The area below Z = -0.58 is 28.10%.
      e.   The area below Z = +3.30 is 99.95% (49.95% + 50% = 99.95%).

4.   a.   The number of cases above Z = +1.71 is 436 (0.0436 × 10,000 = 436).
      b.   The number of cases above Z = -0.83 is 7,967 (0.7967 × 10,000 = 7,967).
      c.   The number of cases above Z = -0.045 is 5,199 (0.5199 × 10,000 = 5,199).
      d.   The number of cases above Z = -0.58 is 7,190 (0.7190 × 10,000 = 7,190).
      e.   The number of cases above Z = +3.30 is 5 (0.0005 × 10,000 = 5).

      a.   The number of cases below Z = +1.71 is 9,564 (0.9564 × 10,000 = 9,564).
      b.   The number of cases below Z = -0.83 is 2,033 (0.2033 × 10,000 = 2,033).
      c.   The number of cases below Z = -0.045 is 4,801 (0.4801 × 10,000 = 4,801).
      d.   The number of cases below Z = -0.58 is 2,810 (0.2810 × 10,000 = 2,810).
      e.   The number of cases below Z = +3.30 is 9,995 (0.9995 × 10,000 = 9,995).

5. $$Z = \frac{X_i - \overline{X}}{s} = \frac{30000 - 27654}{4569} = +0.51$$

The area between the mean salary and a salary of \$30,000 is 19.50%. 69.50% (50+19.5) of the sample fall below a starting salary of \$30,000.

6. $$Z_{Bill} = \frac{X_i - \overline{X}}{s} = \frac{25000 - 27654}{4569} = -0.58$$

$$Z_{Jeff} = \frac{X_i - \overline{X}}{s} = \frac{33500 - 27654}{4569} = +1.28$$

Bill has a Z score of -0.58; Jeff has a Z score of +1.28. The corresponding areas are 21.90% and 39.97%, respectively. The percentage of respondents who have salaries between them is 61.87 (21.9+39.97).

7. $$Z = \frac{X_i - \overline{X}}{s} = \frac{16 - 17.2}{2.3} = -0.52$$

He/she is 0.52 standard deviations below the average years of education. Therefore, 19.85% of the sample is between this score and the mean.

8. $$Z = \frac{X_i - \overline{X}}{s} = \frac{14 - 19.8}{1.3} = -4.46$$

Sue is 4.46 standard deviation units below the mean age in her class. Less than 0.01% of the class is younger than she.

9. $$Z_{18.5} = \frac{X_i - \overline{X}}{s} = \frac{18.5 - 19.8}{1.3} = -1.0 \qquad Z_{21.1} = \frac{X_i - \overline{X}}{s} = \frac{21.1 - 19.8}{1.3} = +1.0$$

If you add and subtract one standard deviation away from the mean, you will get 18.5 and 21.1. Because we know that 68.26% of a sample falls within one standard deviation of the mean, we know that 68.26% of the class are between these two ages.

10. $$Z_{70} = \frac{X_i - \overline{X}}{s} = \frac{70 - 76}{5.9} = -1.02 \qquad Z_{89} = \frac{X_i - \overline{X}}{s} = \frac{89 - 76}{5.9} = +2.20$$

34.61% of the class scored between the mean (76) and a score of 70. 48.61% of the class scored between the mean (76) and a score of 89. Therefore, 83.22% (34.61+48.61) of the class received either a C or a B on the test.

11. $$Z = \frac{X_i - \overline{X}}{s} = \frac{92 - 76}{5.9} = +2.71$$

A grade of 92 has a Z score of +2.71, which means that Sue's test grade was 2.71 standard deviations above the class mean. Or, Sue is 49.66% above the average grade and 99.66% above the entire class. Only 0.34% of the class scored higher than Sue on the exam.

12. $$Z = \frac{X_i - \overline{X}}{s} = \frac{67 - 76}{5.9} = -1.53$$

Jeff's grade is 1.53 standard deviation units below the mean, which means that 43.57% of the class scored between his score and the mean. Because 49.66% of the scores were between the mean and Sue's score, the area between Jeff and Sue's scores is 93.23% (43.57+49.66).

13. $$Z = \frac{X_i - \overline{X}}{s} = \frac{3 - 2.3}{1.7} = +0.41$$

The Z score associated with three children is +0.41; the area beyond 0.41 (column c) is 0.3409. Therefore, 34.09% of the sample have more than three children.

14. $$Z = \frac{X_i - \overline{X}}{s} = \frac{1.9 - 2.3}{1.7} = -0.24$$

The Z score associated with 1.9 is -0.24. The corresponding area between the mean (2.3) and score (column b) is 0.0948. Therefore, the probability of randomly selecting a person from the sample with between 1.9 and 2.3 children is 0.0948.

15.

| | Former Office | Z | New Office | Z |
|---|---|---|---|---|
| Donna | 12.3 | 0.38 | 9.5 | -0.15 |
| Jim | 10.8 | 0.09 | 14.8 | 0.85 |
| Judith | 3.0 | -1.38 | 6.4 | -0.74 |
| Jose | 14.5 | 0.79 | 11.5 | 0.23 |
| Robert | 2.5 | -1.47 | 4.7 | -1.06 |

Donna seems to have cut down her commute time the most, from being 0.38 standard deviations above the mean, to being 0.15 standard deviation units below the mean. Robert seems to still be the closest to the office, though.

16. The lowest SAT score Bill can earn is 850, which is 1100 minus one standard deviation (1100-250=850).

17. $$Z = \frac{X_i - \overline{X}}{s} = \frac{1230 - 1100}{250} = +0.52$$

Ruth is 0.52 standard deviation units above the mean, which indicates that 19.85% of the sample is between her score and the mean and 30.15% of the sample has a higher score. Therefore, 69.85% of the sample fell below her score.

18. $$Z = \frac{X_i - \overline{X}}{s} = \frac{1150 - 1100}{250} = +0.20$$

Carl has a Z score of +0.20, which indicates that 7.93% of the sample is between his score and the mean. 19.85% of the sample is between Ruth's score and the mean. Thus, the area between their two scores is 11.92% (19.85-7.93).

19. Ruth's Z score was +0.52. The area above Ruth's score is 0.3015, which indicates that the probability of someone scoring above Ruth on the SAT is 0.3015. Carl's Z score was +0.20. The area above Carl's score is 0.4207, which indicates that the probability of someone scoring above Carl on the SAT is 0.4207.

## SPSS or MicroCase WORK PROBLEMS

1.  The mean hours of television watched per day by the sample is 2.87 (s=2.793); when converted to z scores, though, the mean is zero and the standard deviation is 1.0. This enables us to then compare scores to the mean. A positive z score indicates that the respondent watches more TV per day than the mean, whereas a negative z score indicates the respondent watches less TV than the mean.

**Descriptive Statistics**

|  | N | Minimum | Maximum | Mean | Std. Deviation |
|---|---|---|---|---|---|
| HOURS PER DAY WATCHING TV | 673 | 0 | 24 | 2.96 | 2.346 |
| Zscore: TVHOURS | 673 | -1.26267 | 8.96533 | .0000000 | 1.00000000 |
| Valid N (listwise) | 673 | | | | |

2.  There are many missing values for the variable "*tvhours*" in the dataset supplied with the textbook. (In the "Data Editor" window of SPSS the missing cases have just a dot in the cell.) In the first 20 cases, there are only seven non-missing cases. The first individual watches 6 hours of TV per day. Given that the overall mean on *tvhours* is 2.96, she has a positive Z score, indicating that she watches more television than average. A person who watches two hours of television a day (as in cases 4, 14, and 19) has a lower Z score (-0.41), indicating she/he is much closer to the overall mean of 2.96. Case 2 watches one hour of television a day; this puts case 2 over three-quarters of a standard deviation below the mean hours watched by the sample. Case 16 watches four hours of TV a day, or nearly one-half of a standard deviation above the mean. (See the table below.)

| CASE NUMBER | RAW SCORE (*TVHOURS*) | Z SCORE (*ZTVHOURS*) | AREA BETWEEN MEAN & Z | AREA ABOVE Z | AREA BELOW Z |
|---|---|---|---|---|---|
| 1 | 6 | 1.29 | 0.4015 | 0.0985 | 0.9015 |
| 2 | 1 | -0.84 | 0.2996 | 0.7996 | 0.2004 |
| 4 | 2 | -0.41 | 0.1591 | 0.6591 | 0.3409 |
| 13 | 6 | 1.29 | 0.4015 | 0.0985 | 0.9015 |
| 14 | 2 | -0.41 | 0.1591 | 0.6591 | 0.3409 |
| 16 | 4 | 0.44 | 0.1700 | 0.3300 | 0.6700 |
| 19 | 2 | -0.41 | 0.1591 | 0.6591 | 0.3409 |

3.     In column 4 of the table above, the area between the mean and each z score is given (these numbers were found in Appendix A in the back of your textbook). For the first and thirteenth respondents, who watch six hours of TV a day, 40.15% of the sample falls between their scores and the mean. If someone watches four hours of television per day, 17.00% of the sample falls between her/his score and the mean hours of television watched per day. For respondents watching two hours of TV a day, 15.91% of the sample falls between their scores and the mean. Last, for someone who watches only one hour a day, 29.96% of the sample falls between their scores and the mean.

4.     In column 4 of the table above, the area above each z score is given (these numbers were found in Appendix A in the back of your textbook). For the first and thirteenth respondents, who watch six hours of TV a day, 9.85% of the sample falls above their scores. If someone watches four hours of television per day, 33.00% of the sample falls above her/his score. For respondents watching two hours of TV a day, 65.91% of the sample falls above their scores. Last, for someone who watches only one hour a day, 79.96% of the sample falls above their scores.

4.     In column 4 of the table above, the area below each z score is given (these numbers were found in Appendix A in the back of your textbook). For the first and thirteenth respondents, who watch six hours of TV a day, 90.15% of the sample falls below their scores. If someone watches four hours of television per day, 67.00% of the sample falls below her/his score. For respondents watching two hours of TV a day, 34.09% of the sample falls below their scores. Last, for someone who watches only one hour a day, 20.04% of the sample falls below their scores.

# CHAPTER SEVEN
## Introduction to Inferential Statistics, the Sampling Distribution, and Estimation

*Learning Objectives: By the end of this chapter, you will be able to:*

1.  *Explain the purpose of inferential statistics in terms of generalizing from a sample to a population.*
2.  *Explain the principle of random sampling and these key terms: population, sample, parameter, statistic, representative, EPSEM.*
3.  *Differentiate among the sampling distribution, the sample, and the population.*
4.  *Explain the two theorems presented.*
5.  *Explain the logic of estimation and the role of the sample, sampling distribution, and population.*
6.  *Define and explain the concepts of bias and efficiency.*
7.  *Construct and interpret confidence intervals for sample means and sample proportions.*

## KEY TERMS

- ✓ Alpha ($\alpha$) (p. 158)
- ✓ Bias (p. 155)
- ✓ Central Limit Theorem (p. 151)
- ✓ Confidence Interval (p. 155)
- ✓ Confidence Level (p. 158)
- ✓ Efficiency (p. 156)
- ✓ EPSEM (p. 147)
- ✓ $\mu$ (p. 154)
- ✓ $\mu_0$ (p. 154)
- ✓ $\mu_p$ (p. 154)
- ✓ Parameter (p. 146)
- ✓ $P_s$ (p. 154)
- ✓ $P_u$ (p. 154)
- ✓ Representative (p. 147)
- ✓ Sampling Distribution (p. 149)
- ✓ Simple Random Sample (p. 147)
- ✓ Standard Error of the Mean (p. 151)

# CHAPTER SUMMARY

In the first six chapters of your text, you learned descriptive statistics: ways to describe the distribution of your sample on some given variable. Now, we move onto inferential statistics: ways in which we can <u>infer</u> our sample results to the rest of the population. To do so, we make use of the concept of a sampling distribution to estimate population parameters from sample statistics.

# CHAPTER OUTLINE

**7.1   Introduction.**

**7.2   Probability Sampling.**

- Two important and very distinct concepts are (1) samples and (2) populations. Populations are the entire body of who or what you are studying. If you are interested in women of childbearing age in the U.S., then your population is <u>all</u> women between the ages of 13 and 50 in the United States. Because it is virtually impossible to reach all of these women, you need to <u>sample</u> this population. That is, you need to randomly select a subset of the population that will represent all women in the U.S. between the ages of 13 and 50.

- In order to do so you must use a probability sampling technique, also called "random sampling." The basic idea behind probability sampling is that every single person in your population has an equal probability of being selected for your sample, thus assuring that your sample is representative of the entire population. The goal of probability sampling is to produce a sample that is representative of the population from which it was drawn. A sample is considered representative when it reproduces important population characteristics.

- For example, let's say that you're interested in height and shoe size, and you decide to do a study on the relationship between these two variables. In order to find a quick sample, you decide to walk to the gym where the men's basketball team practices. As each player leaves the court, you ask each one his height and his shoe size. You then take this data home and find that everyone is <u>very</u> tall and has <u>very</u> large feet. In fact, you go so far as to conclude that height and shoe size are related: if you have large feet, you are tall, and if you are tall, you have large feet.

- What's wrong with your data? Well, your sample is not representative of the whole population. In fact, it does not even represent all basketball players. It is only using one population: the basketball players in that particular gym at that time. Because of the non-randomness of your sample, you cannot infer your results to anyone besides the basketball players in the gym. Therefore, although the relationship between these two variables may be true, the results are not generalizable onto the population.

- Thus, social scientists ensure that they can infer their results by utilizing probability sampling techniques. Unfortunately, a representative sample can never be guaranteed, but we can maximize our chances of drawing a representative sample by following the principle of EPSEM (Equal Probability of Selection Method). The most basic EPSEM sampling technique is called "simple random sampling," which is described more fully in your textbook.

## 7.3 The Sampling Distribution.

- Describing a sample is fine and good, and in fact it is an important first step in analyzing data, but the main point in social science research is to make a statement about the *entire* population, not just the sample. Therefore, inferential statistics are a mainstay to statistical analysis.
- In order to make these inferences, social scientists make use of the sampling distribution, which is a theoretical notion that if you could take an infinite number of samples, of let's say 100 respondents each (N=100), and you plotted each sample's mean of height, the sampling distribution of means would represent a normal curve. That is, we are making the assumption, with the use of this theoretical sampling distribution, that if enough samples are collected, the distribution of these sample means will be shaped normally. Stated in yet another way, we are saying that we know that in theory that an infinite number of samples of all of the same size will produce a distribution of means that looks like a normal distribution. Because we know this to be true, then, we can assume that if we take only one sample of that same size, it will be within the range of the normal distribution and thus be representative of the population. But why in the world would anyone ever care about such things, you ask yourself. Because without this assumption, we cannot make inferences about the population. We need this theoretical framework in order to say that our sample results can be inferred to the population, which, as we stated before, is what we strive for.
- Although the sampling distribution is entirely theoretical, we can describe the distribution by making use of the two theorems presented in your textbook. The two theorems are:
  1. If repeated random samples of size N are drawn from a normal population with mean $\mu$ and standard deviation $\sigma$, then the sampling distribution of sample means will be normal with a mean $\mu$ and a standard deviation of $\sigma/\sqrt{n}$.
  2. If repeated random samples of size N are drawn from any population, with mean $\mu$ and standard deviation $\sigma$, then, as N becomes large, the sampling distribution of sample means will approach normality, with mean $\mu$ and standard deviation $\sigma/\sqrt{n}$.

- Both theorems describe the shape (normal), center ($\mu$), and dispersion ($\sigma/\sqrt{n}$) of the theoretical sampling distribution. (The dispersion statistic for the sampling distribution has a special name; it's called the standard error.) The second theorem, which is called the Central Limit Theorem, removes the first theorem's requirement of a normal population. Whenever the sample size (N) is "large" (a good "rule of thumb" is N≥100), we can assume that the sampling distribution is normal and make use of the Central Limit Theorem to link the sample distribution to the population distribution through the sampling distribution.

## 7.4   The Sampling Distribution: An Additional Example.

## 7.5   Symbols and Terminology.
- For samples, we use $\bar{X}$ to represent the mean, $S$ to represent the standard deviation, and $P_s$ to represent the proportion.
- For populations, we use $M$ or $\mu$ to represent the mean, $\sigma$ to represent the standard deviation, and $P_u$ to represent the proportion.
- For sampling distributions of sample means, we use $\mu_{\bar{x}}$ to represent the mean and $\sigma_{\bar{x}}$ to represent the standard deviation.
- For sampling distributions of sample proportions, we use $\mu_p$ to represent the mean and $\sigma_p$ to represent the standard deviation.

## 7.6   Introduction to Estimation.

## 7.7   Bias and Efficiency.
- Estimation procedures attempt to estimate the *population* values by making use of our *sample* values. In order to do so, our estimators must be both unbiased (the mean of the sampling distribution is equal to the population mean) and efficient (the extent to which the sampling distribution is clustered around the mean). Basically this means that we are <u>assuming</u> our estimator to be unbiased. We know that as our sample size increases, so too does the efficiency of our estimator.

## 7.8   Estimation Procedures: Introduction.

## 7.9   Interval Estimation Procedures for Sample Means (Large Samples).
- Your textbook focuses on one main type of estimate, confidence intervals. Confidence intervals use a single-sample estimate to find a <u>range</u> in which we would estimate the population value to fall. This range is the confidence interval, in which we are fairly confident that the population value will fall between.

- Here's a nice example of how confidence intervals work (adapted from Blalock, 1979[1]). Let's say you have a block of wood that weighs 8 ounces. The weight of wood, being the breathable material that it is, though, is not constant. So you weigh it a second time and it weighs 7.9 ounces. A third weighing indicates that it is 8.2 ounces; a fourth is 8.3; and again at 7.8; and so on. After you weigh this wood several more times, you can get an average weight of the piece of wood, let's say 8.04 ounces, plus or minus 0.16 ounces. Therefore you are admitting that you don't know exactly how much this piece of wood weighs, but you're fairly confident that its weight is in this range, or confidence interval. Confidence intervals are used to estimate a range of population values from a sample value.

- How confident we are that this range truly does contain the population value depends largely on the amount of error we can tolerate. We decide beforehand how often we are going to be right. Do we want to be correct 50% of the time? No, that's too low. How about 75% of the time? Well, that means we're going to be wrong 25% of the time--still too low. Ok, how about 95% of the time? Well, that seems pretty reasonable. Let's say we're going to be correct about estimating the population value 95% of the time. That means we're going to be wrong only 5% of the time. Those odds are good enough for Las Vegas! Making this decision, the probability of being wrong, is called choosing an alpha level. If we set our alpha level at 0.05, or 5%, then we are going to be wrong only 5% of the time. We could set our alpha at any level, but 0.05 is used most often in the social sciences. The alpha level is tied to the confidence level. If the alpha level is how "wrong" we're willing to be, then the confidence level is how "right" or confident in our results we want to be. For example, when the alpha level is 0.05, the confidence level is 95% (that is, we're willing to be wrong 5% of the time and hope to be right 95% of the time). Similarly, when the alpha level is 0.01, the confidence level is 99%.

- To calculate a confidence interval for interval-ratio data (a sample mean), we use the following equation:

$$\text{confidence interval} (c.i.) = \overline{X} \pm Z\left(\frac{\sigma}{\sqrt{N}}\right)$$

- This equation estimates the population confidence interval, given the sample mean and the standard deviation of the sampling distribution, which in turn makes use of the population standard deviation ($\sigma$). Because we rarely know $\sigma$, we estimate it with the standard deviation of the sample (s), and use the following equation:

$$c.i. = \overline{X} \pm Z\left(\frac{s}{\sqrt{N-1}}\right)$$

---

[1] Blalock, Hubert M, Jr. 1979. *Social Statistics*. New York: McGraw-Hill, Inc.

- This equation allows us to estimate the population confidence interval with the sample mean and the sample standard deviation. What we find is a range of values in which we are 95% confident that the population mean will fall between, knowing, though, there is a 5% chance that the population mean does <u>not</u> fall within this range.

## 7.10  Interval Estimation Procedures for Sample Proportions (Large Samples).

- We can also estimate the population value for nominal data by using proportions. Let's say 9% of our sample of 250 respondents are black, and we want to estimate the proportion of blacks in the population. Using the following equation, we can get this estimate:

$$c.i. = P_s \pm Z \sqrt{\frac{P_u(1-P_u)}{N}}$$

- Our sample proportion ($P_s$) is 0.09, and Z is 1.96 when alpha is 0.05. We set the population proportion ($P_u$) to be 0.5 (because this is the most conservative estimate). We can then solve this equation:

$$c.i. = 0.09 \pm 1.96 \sqrt{\frac{0.5(1-0.5)}{250}} = 0.09 \pm 0.06$$

- Therefore, we can infer from our sample that the proportion of blacks in the population is between 0.03 and 0.15.

## 7.11  A Summary of the Computation of Confidence Intervals.

## 7.12  Controlling the Width of Interval Estimates.

- The width of the confidence interval is dependent on two things: the alpha level and the sample size. The more errors you are willing to make (the higher the alpha level and the lower the confidence level), the smaller your interval width. Thus if you are willing to be wrong 10% of the time, your interval width will be smaller than if you are willing to be wrong only 1% of the time.
- Also, increased sample size will decrease your interval width. As your sample size increases, you are collecting a wider range of data on the population, and thus you are able to make a better prediction of the population value.

**MULTIPLE-CHOICE QUESTIONS**

1.      Parameters are _____.
        a. statistical methods of inferential statistics
        b. the distance between descriptive and inferential statistics
        c. sampling techniques for inferential statistics
        d. generalized population characteristics

        *Learning Objective: 2*

2.      EPSEM is the acronym for_____.
        a. Equal Probability of Selection Method
        b. Equal Probability of Service Employment Method
        c. Earnest Plying of Selected Entry Method
        d. Every Person Serves Equally Method

        *Learning Objective: 2*

3.      A sample is representative if it _____.
        a. falls within the sampling distribution
        b. is greater than 1000
        c. reproduces the major characteristic of the population
        d. all of these choices

        *Learning Objective: 2*

4.      A simple random sample is a method in which _____.
        a. every case has an equal chance of being chosen
        b. every other case is chosen for the sample
        c. the sample predicts the population
        d. the population is used to estimate the parameter

        *Learning Objective: 2*

5.      The central concept to inferential statistics is _____.
        a. EPSEM
        b. stratified sampling techniques
        c. sampling distribution
        d. population parameters

        *Learning Objective: 5*

6. The three distributions involved in every application of inferential statistics are
_____.
   a. sample proportion, population proportion, sampling mean
   b. random sampling, sample distribution, sampling probability
   c. sample distribution, population distribution, sampling distribution
   d. EPSEM, sample means, sampling distribution

*Learning Objective: 3*


7. The sampling distribution is _____.
   a. known
   b. theoretical
   c. empirical
   d. random

*Learning Objective: 3*


8. The sampling distribution is _____.
   a. probabilistic
   b. improbable
   c. real
   d. unknown

*Learning Objective: 3*


9. The sample and the population distributions are both _____, but only the sample is
_____.
   a. unknown, empirical
   b. known, empirical
   c. probabilistic, theoretical
   d. empirical, known

*Learning Objective: 3*


10. The standard error of the mean is the _____.
   a. center of the sampling distribution
   b. standard deviation of the sampling distribution
   c. standard deviation of the population
   d. skew of the sampling distribution

*Learning Objective: 3*

11. The Central Limit Theorem describes _____.
    a. the shape, center, and spread of the population distribution
    b. the shape, center, and spread of the sample distribution
    c. the shape, center and spread of the sampling distribution
    d. the normality of the sampling distribution

    *Learning Objective: 4*


12. As sample size increases, the spread of the sampling distribution _____.
    a. decreases
    b. increases
    c. stays the same
    d. increases to some point and then decreases

    *Learning Objective: 4*


13. A confidence interval is _____.
    a. an estimate of the sample distribution
    b. an estimate of the sample size
    c. an estimate of the population size
    d. an estimate of the population value

    *Learning Objective: 7*


14. A good estimator should be _____ and _____.
    a. biased, efficient
    b. biased, evasive
    c. unbiased, inefficient
    d. unbiased, efficient

    *Learning Objective: 6*


15. A statistic is unbiased if _____.
    a. the mean of the sampling distribution is equal to the sample
    b. the sample size is larger than 50
    c. the mean of the sample is equal to the mean of the population
    d. the mean of the sampling distribution is equal to the population

    *Learning Objective: 6*

16. Efficiency of a measure indicates _____.
    a. the degree to which the population mean is equal to the standard deviation of the sample means
    b. the degree to which the sample outcomes are clustered around the mean of the sampling distribution
    c. the distance between the population value and the sample value
    d. the population value can be found quickly and easily

    *Learning Objective: 6*


17. Alpha levels are _____.
    a. the amount of error in a sample
    b. the probability of the population
    c. the probability of error
    d. the Greek letter for population variance

    *Learning Objective: 5*


18. The larger the sample size, the _____ the confidence interval.
    a. smaller
    b. wider
    c. more continuous
    d. greater

    *Learning Objective: 7*


19. The higher the confidence level, the _____ the interval width.
    a. smaller
    b. wider
    c. more continuous
    d. more precise

    *Learning Objective: 7*


20. A alpha level of 0.05 has a Z score of _____.
    a. $\pm 3.25$
    b. $\pm 2.58$
    c. $\pm 1.96$
    d. $\pm 1.65$

    *Learning Objective: 5*

## WORK PROBLEMS

1. Construct a confidence interval for each of the sample outcomes below at the 0.01 and 0.05 levels. Discuss the differences between the 95% and 99% confidence intervals.

   a. $\overline{X} = 34.6$  b. $\overline{X} = 68$
   s=12.3   s=25
   N=187   N=540

   c. $\overline{X} = 164.8$  d. $\overline{X} = 76$
   s= 83.2   s= 9.6
   N=150   N=100

   e. Ps=0.35  f. Ps=0.56
   N =112   N =456

   *Learning Objective: 7*

2. Find the corresponding Z scores for the confidence levels below.

   | Confidence Level | Alpha | Z | Area Beyond Z Score |
   | --- | --- | --- | --- |
   | 90% | 0.10 | 0.0505 | 1.64 |
   | 95% | | | |
   | 99% | | | |
   | 80% | | | |
   | 75% | | | |

   *Learning Objective: 7*

3. In a study of attitudes toward smoking, a researcher found that 75 of his 238 respondents smoke. Estimate the proportion of smokers in the population at an alpha of 0.05.

   *Learning Objective: 7*

4. Using the smoking attitudinal data from the previous problem, estimate the proportion at an alpha of 0.10.

   *Learning Objective: 7*

134

5.      Using the smoking attitudinal data from problem 3, estimate the proportion at an alpha of 0.01.

*Learning Objective: 7*


6.      Using the smoking attitudinal data from problem 3, estimate the proportion at an alpha of 0.05, only assume a sample size of 2380.

*Learning Objective: 7*


7.      Describe and explain the changes in the estimates you calculated for problems 3 through 6.

*Learning Objective:  7*


8.      In a sample of students at a university (N=1200), the grade point average was 3.2, with a standard deviation of 0.9.  Estimate the grade point average for the entire university, at a 90% confidence level.

*Learning Objective: 7*


9.      The average grade of a sample of 125 students in a statistics course at a university was 78.5, with a standard deviation of 15.7.  Estimate the population mean for an alpha of 0.05.

*Learning Objective: 7*


10.     Out of a sample of 325, 175 respondents stated that they were in favor of women's rights. Estimate the proportion of the population that favors women's rights, at the 95% confidence interval.

*Learning Objective: 7*


11.     A researcher found in a sample of 238 households that the average income is $31,457, with a standard deviation of $15,722.  Estimate the population income at a 99% confidence level.

*Learning Objective: 7*

12. In a sample of 125 inmates at a local prison, the mean sentence was 8 years, with a standard deviation of 3 years. Estimate the mean sentence for the population of all prison inmates at the 0.05 alpha level.

*Learning Objective: 7*

13. In a similar study the same mean and standard deviation were found, but the sample size was 1200. How does this change the results. Explain.

*Learning Objective: 7*

14. In a sample of 450 graduating seniors at Lowdown College, 289 have found jobs. Estimate the proportion of the population who have jobs. (Use alpha = 0.05.)

*Learning Objective: 7*

15. In a study of women's health, it was found that on average, women went to see a physician for a regular check up 1.3 times per year (s=4.3). From this sample of 300 women, estimate the number of physician visits for the entire population, with only a 1% chance of error.

*Learning Objective: 7*

16. A survey of the men's basketball teams at some large universities (N=135) found that the average height was 6'2" (74 inches), plus or minus 6 inches. Estimate the mean height of all male basketball players. (Use alpha = 0.05.)

*Learning Objective: 7*

17. In a survey of attitudes toward racially mixed marriages, a researcher found that the majority of her sample (235 out of 300) did not have a problem with them. Given a 90% confidence level, what proportion of the population has the same attitude?

*Learning Objective: 7*

18. A drug company is testing their new allergy medicine. Of the 1750 people being tested, 123 had negative reactions. Because they are very concerned about the health and well-being of their customers, they want to be very sure of their results. At a 99% confidence level, estimate the proportion of negative reactions in the population.

*Learning Objective: 7*

## SPSS or MicroCase WORK PROBLEMS

The following work problems can be completed using either SPSS or MicroCase. The solutions will be the same regardless of which one you use. Your instructor may have a preference, so be sure to check with him/her prior to beginning these problems.

1. If we assume that the entire GSS 2006 dataset is a population, then we can measure the effect of repeated random samples and sample size by taking random samples of this "population" dataset. Select four random samples from the dataset that range in size: (1) 10% of the "population," (2) 25% of the "population," (3) 50% of the "population," and (4) 75% of the "population." After you've selected each one, find the mean age for each sample. Describe what happens as the sample size increases.

Make sure to follow all of the analysis steps described below.

Here are the steps for selecting random samples: Select the "Data" menu and choose the "Select Cases" option. In the "Select Cases" window, click the "Random sample of cases" option and then click on the "Sample" button. Select the "Approximately __% of all cases" option (it may already be selected); type 10 in the box for a 10% sample (type 25 for a 25% sample, etc.). Click the "Continue" button. Back in the "Select Cases" window, make sure that the "Unselected Cases Are ... Filtered" option is checked. Click the "OK" button. Run the appropriate analysis (such as requesting the mean age with the Descriptives command). It is very important that you go back to the original "population" between each sampling. To return to the original full "population" select the "Data" menu, choose the "Select Cases" option, click the "All cases" option, and click the "OK" button. Now you can request the next sample.

In the next problem you're asked to obtain confidence intervals for each of your samples. It would be easiest to request the mean and confidence interval for each sample at the same time. To do so, use the following procedure: Select the "Analyze" menu, select "Descriptive Statistics" and then "Explore." In the "Explore" window, enter the *age* variable in the "Dependent List" box. Click the "Statistics" button; check the "Descriptives" option and enter the confidence level (for example, 95) in the little box ("Confidence Interval for Mean: __%"). Click the "Continue" button. Back in the "Explore" window make sure that the "Statistics" option is checked under "Display." Click the "OK" button. You'll want to do this in the beginning of your analysis for the "population" value and repeat it each time you take a sample. That way you'll have the means and 95% confidence intervals for each sample. For the last sample, the 75% sample, request three different confidence intervals: 95%, 99%, and 90%.

2.    In the previous problem you described what happens to the various means as the sample sizes increased. Now, using the output produced for problem 1, discuss what happens to the confidence intervals as the sample sizes increase. Is the "population" value for the mean age contained within the confidence interval? For the last 75% sample, discuss what happens to the confidence intervals as the confidence level changes.

3.    Using the method described above, request a 95% confidence interval for *childs*.

4.    Using the method described above, request a 99% confidence interval for *realinc*.

## ANSWER KEY

## MULTIPLE-CHOICE QUESTIONS

| | | | | | | |
|---|---|---|---|---|---|---|
| 1. | d | (p. 146) | 11. | c | (p. 151) | |
| 2. | a | (p. 147) | 12. | a | (p. 149-151) | |
| 3. | c | (p. 147) | 13. | d | (p. 155) | |
| 4. | a | (p. 147) | 14. | d | (p. 155) | |
| 5. | c | (p. 146) | 15. | d | (p. 155-156) | |
| 6. | c | (p. 149) | 16. | b | (p. 156-158) | |
| 7. | b | (p. 149) | 17. | c | (p. 158) | |
| 8. | a | (p. 149) | 18. | a | (p. 169-171) | |
| 9. | d | (p. 149) | 19. | b | (p. 169-171) | |
| 10. | b | (p. 151) | 20. | c | (p. 158-160) | |

## WORK PROBLEMS

1.    Parts a through e use the following two formulas to calculate the confidence intervals:

$$c.i. \text{(for mean)} = \overline{X} \pm Z\left(\frac{s}{\sqrt{N-1}}\right) \quad \text{or} \quad c.i. \text{(for proportion)} = P_s \pm Z\left(\sqrt{\frac{P_u(1-P_u)}{N}}\right)$$

a.    $95\% : 34.6 \pm 1.96\left(\frac{12.3}{\sqrt{187-1}}\right) = 34.6 \pm 1.77$

$99\%: \quad 34.6 \pm 2.58 \left( \dfrac{12.3}{\sqrt{187-1}} \right) = 34.6 \pm 2.33$

b. $\quad 95\%: \quad 68 \pm 1.96 \left( \dfrac{25}{\sqrt{540-1}} \right) = 68 \pm 2.11$

$\quad 99\%: \quad 68 \pm 2.58 \left( \dfrac{25}{\sqrt{540-1}} \right) = 68 \pm 2.78$

c. $\quad 95\%: 164.8 \pm 1.96 \left( \dfrac{83.2}{\sqrt{150-1}} \right) = 164.8 \pm 13.36$

$\quad 99\%: 164.8 \pm 2.58 \left( \dfrac{83.2}{\sqrt{150-1}} \right) = 164.8 \pm 17.59$

d. $\quad 95\%: \quad 76 \pm 1.96 \left( \dfrac{9.6}{\sqrt{100-1}} \right) = 76 \pm 1.89$

$\quad 99\%: \quad 76 \pm 2.58 \left( \dfrac{9.6}{\sqrt{100-1}} \right) = 76 \pm 2.49$

e. $\quad 95\%: 0.35 \pm 1.96 \left( \sqrt{\dfrac{(0.5)(1-0.5)}{112}} \right) = 0.35 \pm 0.09$

$\quad 99\%: 0.35 \pm 2.58 \left( \sqrt{\dfrac{(0.5)(1-0.5)}{112}} \right) = 0.35 \pm 0.12$

f. $\quad 95\%: 0.56 \pm 1.96 \left( \sqrt{\dfrac{(0.5)(1-0.5)}{456}} \right) = 0.56 \pm 0.05$

$\quad 99\%: 0.56 \pm 2.58 \left( \sqrt{\dfrac{(0.5)(1-0.5)}{456}} \right) = 0.56 \pm 0.06$

2.

| Confidence Level | Alpha | Z | Area Beyond Z Score |
|---|---|---|---|
| 90% | 0.10 | 0.0505 | 1.64 |
| 95% | 0.05 | 0.0250 | 1.96 |
| 99% | 0.01 | 0.0049 | 2.58 |
| 80% | 0.20 | 0.0985 | 1.29 |
| 75% | 0.25 | 0.1230 | 1.16 |

3. $$c.i. = P_s \pm Z\left(\sqrt{\frac{(0.5)(1-0.5)}{N}}\right) = \frac{75}{238} \pm 1.96\left(\sqrt{\frac{(0.5)(0.5)}{238}}\right) = 0.32 \pm 0.06$$

Because this is nominal data, you first need to find the proportion of smokers in the sample (75/238 = 0.32). The estimate of the proportion of smokers in the population is 0.32 ± 0.06. Or, I am 95% confident that the total proportion of smokers in the population is between 0.26 and 0.38.

4. $$c.i. = P_s \pm Z\left(\sqrt{\frac{(0.5)(1-0.5)}{N}}\right) = \frac{75}{238} \pm 1.64\left(\sqrt{\frac{(0.5)(0.5)}{238}}\right) = 0.32 \pm 0.05$$

The estimate of the proportion of smokers in the population is 0.32 ± 0.05. Or, I am 90% confident that the total proportion of smokers in the population is between 0.27 and 0.37.

5. $$c.i. = P_s \pm Z\left(\sqrt{\frac{(0.5)(1-0.5)}{N}}\right) = \frac{75}{238} \pm 2.58\left(\sqrt{\frac{(0.5)(0.5)}{238}}\right) = 0.32 \pm 0.08$$

The estimate of the proportion of smokers in the population is 0.32 ± 0.08. Or, I am 99% confident that the total proportion of smokers in the population is between 0.24 and 0.40.

6. $$c.i. = P_s \pm Z\left(\sqrt{\frac{(0.5)(1-0.5)}{N}}\right) = \frac{75}{238} \pm 1.96\left(\sqrt{\frac{(0.5)(0.5)}{2380}}\right) = 0.32 \pm 0.02$$

The estimate of the proportion of smokers in the population is 0.32 ± 0.02. Or, I am 95% confident that the total proportion of smokers in the population is between 0.30 and 0.34.

7. As the confidence level increased from 90% to 95% and 99%, the width of the confidence intervals also increased.

8. $$c.i. = \overline{X} \pm Z\left(\frac{s}{\sqrt{N-1}}\right) = 3.2 \pm 1.65\left(\frac{0.9}{\sqrt{1200-1}}\right) = 3.2 \pm 0.05$$

Because GPA is an interval-ratio variable, we want to estimate the population mean, which is 3.2 ± 0.05. I am 90% confident that the mean GPA in the population is between 3.15 and 3.25. The interval width is small because of the large sample size.

9. $$c.i. = \overline{X} \pm Z\left(\frac{s}{\sqrt{N-1}}\right) = 78.5 \pm 1.96\left(\frac{15.7}{\sqrt{125-1}}\right) = 78.5 \pm 2.76$$

I am 95% confident that the mean grade in the population is between 75.74 and 81.26.

10. $$c.i. = P_s \pm Z\left(\sqrt{\frac{(0.5)(1-0.5)}{N}}\right) = \frac{175}{325} \pm 1.96\left(\sqrt{\frac{(0.5)(0.5)}{325}}\right) = 0.54 \pm 0.06$$

The proportion of sample respondents who are in favor of women's rights is 0.54 (175/325). Therefore, I am 95% confident that between 48% and 60% of the population are in favor of women's rights.

*15.394804318*

11. $$c.i. = \overline{X} \pm Z\left(\frac{s}{\sqrt{N-1}}\right) = 31457 \pm 2.58\left(\frac{15722}{\sqrt{238-1}}\right) = 31457 \pm 2635.60$$

I am 99% confident that the mean income in the population is between \$28,821.40 and \$34,092.60.

*11.13552872566*

12. $$c.i. = \overline{X} \pm Z\left(\frac{s}{\sqrt{N-1}}\right) = 8 \pm 1.96\left(\frac{3}{\sqrt{125-1}}\right) = 8 \pm 0.53$$

Therefore, I am 95% confident that the mean sentence for the population is between 7.47 years and 8.53 years.

13. $$c.i. = \overline{X} \pm Z\left(\frac{s}{\sqrt{N-1}}\right) = 8 \pm 1.96\left(\frac{3}{\sqrt{1199-1}}\right) = 8 \pm 0.17$$

With the increased sample size, we are better able to estimate the value of the population mean, therefore the width of the interval is reduced. I am 95% confident that the mean sentence for the population is between 7.83 years and 8.17 years.

14. $$c.i. = P_s \pm Z\left(\sqrt{\frac{(0.5)(1-0.5)}{N}}\right) = \frac{289}{450} \pm 1.96\left(\sqrt{\frac{(0.5)(0.5)}{450}}\right) = 0.64 \pm 0.05$$

The proportion of graduating seniors who have found employment in the sample is 0.64 (289/450). Therefore, I am 95% confident that between 59% and 69% of the population of graduating seniors have found employment.

15. $$c.i. = \overline{X} \pm Z\left(\frac{s}{\sqrt{N-1}}\right) = 1.3 \pm 2.58\left(\frac{4.3}{\sqrt{300-1}}\right) = 1.3 \pm 0.65$$

I am 99% confident that the population of women sees a physician between 0.65 and 1.95 times per year.

16. $$c.i. = \overline{X} \pm Z\left(\frac{s}{\sqrt{N-1}}\right) = 74 \pm 1.96\left(\frac{6}{\sqrt{135-1}}\right) = 74 \pm 1.02$$

I am 95% confident that the mean height for all basketball players then is between 6'1" and 6'3".

17. $$c.i. = P_s \pm Z\left(\sqrt{\frac{(0.5)(1-0.5)}{N}}\right) = \frac{235}{300} \pm 1.65\left(\sqrt{\frac{(0.5)(0.5)}{300}}\right) = 0.78 \pm 0.05$$

I am 90% confident that between 73% and 83% of the population do not have a problem with racially mixed marriages.

18. $$c.i. = P_s \pm Z\left(\sqrt{\frac{(0.5)(1-0.5)}{N}}\right) = \frac{123}{1750} \pm 2.58\left(\sqrt{\frac{(0.5)(0.5)}{1750}}\right) = 0.07 \pm 0.03$$

I am 99% confident that between 4% and 10% of the population is estimated to have a negative reaction to the medication.

## SPSS or MicroCase WORK PROBLEMS

If you had problems with any of these questions, be sure to re-read the appropriate sections at the end of chapter 6 in your textbook.

1. The "population" mean age is 47.43 years (s=16.798).

**Descriptive Statistics**

|  | N | Minimum | Maximum | Mean | Std. Deviation |
|---|---|---|---|---|---|
| age | 1493 | 18 | 89 | 47.43 | 16.912 |
| Valid N (listwise) | 1493 |  |  |  |  |

In order to save space, I've combined the "Explore" output for the different samples on the next page. I've also removed some of the extra information. Because SPSS takes the random samples differently each time, your means and confidence intervals may be slightly different than mine.

**Descriptives**

| | | | Statistic |
|---|---|---|---|
| 10% sample (N=154) | Mean | | 46.92 |
| | 95% Confidence Interval for Mean | Lower Bound | 44.24 |
| | | Upper Bound | 49.59 |
| 25% sample (N=340) | Mean | | 47.49 |
| | 95% Confidence Interval for Mean | Lower Bound | 45.66 |
| | | Upper Bound | 49.31 |
| 50% sample (N=744) | Mean | | 46.36 |
| | 95% Confidence Interval for Mean | Lower Bound | 45.78 |
| | | Upper Bound | 47.56 |
| 75% sample (N=1116) | Mean | | 47.38 |
| | 95% Confidence Interval for Mean | Lower Bound | 46.38 |
| | | Upper Bound | 48.38 |
| | 99% Confidence Interval for Mean | Lower Bound | 46.07 |
| | | Upper Bound | 48.69 |
| | 90% Confidence Interval for Mean | Lower Bound | 46.55 |
| | | Upper Bound | 48.22 |

Comparing the random samples from each of the samples above, we see that a sample of 10% of the "population" yields a mean age of 46.92 years, compared to 47.49 years when 25% of the "population" is sampled, 46.36 years when 50% of the "population" is sampled, and 47.38 when 75% of the "population" is sampled. Overall all four of these sample means are fairly close to the "population" mean of 47.43. If you correctly sample a population and your sample size is large, then the sample mean will approximate the population mean.

2.  The second computer work problem asked you to compare the 95% confidence intervals for the four samples. Using the information above, we can see that the 95% confidence interval for the 10% sample is between 44.24 and 49.59. The 95% confidence interval for the 25% sample is between 45.66 and 49.31, between 45.78 and 47.56 for the 50% sample, and between 46.38 and 48.38 for the 75% sample. All four of these confidence intervals contain the "population" value of 47.43. (Remember, we were only 95% sure that they would.) Notice how the width of the confidence intervals decreases as sample size increases. This is because larger samples provide better estimates of the population.

We were also asked to compare the 95%, 99%, and 90% confidence intervals for the 75% sample. In order, the 90% confidence interval ranges from 46.55 to 48.22; the 95% confidence interval ranges 46.38 to 48.38; and the 99% confidence interval ranges from 46.07 to 48.69. Notice how the interval width increases with the confidence level. In order for us to be more confident that the interval contains the population value, it must be a wider interval. We have less confidence that our interval contains the population value when it's narrower.

143

3.  The 95% confidence interval ranges from 1.83 to 2.00.  Therefore, we are 95% confident that respondents have between 1.83 and 2.00 children in the population from which the sample was drawn.

**Descriptives**

|  |  |  | Statistic | Std. Error |
|---|---|---|---|---|
| childs | Mean |  | 1.92 | .043 |
|  | 95% Confidence Interval for Mean | Lower Bound | 1.83 |  |
|  |  | Upper Bound | 2.00 |  |

4.  The 99% confidence interval ranges from $30,772.29 to $35,327.52.  Therefore, we are 99% confident that respondents make between  $30,772.29 to $35,327.52 in the population from which the sample was drawn.

**Descriptives**

|  |  |  | Statistic | Std. Error |
|---|---|---|---|---|
| realinc | Mean |  | 33049.90 | 882.908 |
|  | 95% Confidence Interval for Mean | Lower Bound | 30772.29 |  |
|  |  | Upper Bound | 35327.52 |  |

# CHAPTER EIGHT
## Hypothesis Testing I:  The One-Sample Case

_Learning Objectives_: By the end of this chapter, you will be able to:

1.  Explain the logic of hypothesis testing.
2.  Define and explain the conceptual elements involved in hypothesis testing, especially the null hypothesis, the sampling distribution, the alpha level, and the test statistic.
3.  Explain what it means to "reject the null hypothesis" or "fail to reject the null hypothesis."
4.  Identify and cite examples of situations in which one-sample tests of hypothesis are appropriate.
5.  Test the significance of single-sample means and proportions using the five-step model and correctly interpret the results.
6.  Explain the difference between one- and two-tailed tests and specify when each is appropriate.
7.  Define and explain Type I and Type II errors and relate each to the selection of an alpha level.

## KEY TERMS

- ✓ Alpha Level ($\alpha$)  (p. 185)
- ✓ Critical Region (Region of Rejection)  (p. 184)
- ✓ Five-Step Model  (p. 183)
- ✓ Hypothesis Testing  (p. 177)
- ✓ Null Hypothesis ($H_o$)  (p. 183)
- ✓ One-Tailed Test  (p. 187)
- ✓ Research Hypothesis ($H_1$)  (p. 184)
- ✓ Significance Testing  (p. 177)
- ✓ _t (critical)_  (p. 194)
- ✓ Student's _t_ Distribution  (p. 193)
- ✓ _t (obtained)_  (p. 194)
- ✓ Test Statistic  (p. 185)
- ✓ Two-Tailed Test  (p. 187)
- ✓ Type I Error (Alpha Error)  (p. 191)
- ✓ Type II Error (Beta Error)  (p. 192)
- ✓ _Z (critical)_  (p. 184)
- ✓ _Z (obtained)_  (p. 185)

# CHAPTER SUMMARY

In chapter eight of your textbook, you are introduced to the idea of hypothesis testing. This chapter introduces one sample hypothesis testing where we are trying to discern whether or not our sample is really from the population, or from some other population. Sounds strange, doesn't it? But, follow along, and you'll probably agree that this is an important concept.

# CHAPTER OUTLINE

## 8.1 Introduction.

## 8.2 An Overview of Hypothesis Testing.

- Basically it works like this: let's say we are interested in height and shoe size, but we think that basketball players will be much taller and have much larger feet than the rest of the population. In fact, you could consider all basketball players as being a separate population from all other people, non-basketball players. We could take a sample from the population of all basketball players and compare the average height of this sample of the population to the average height of the other population (non-basketball players). Thus we would be comparing the mean height of the sample from the basketball population ($\overline{X}$) to the known non-basketball population mean (:) in order to find out whether or not the population of basketball players is significantly different from the population of non-basketball players. If we find a significant difference, then we decide that our sample of basketball players really is from a different population than the rest of the population.

- Let me stress again: we are trying to decide whether or not our sample of basketball players is from the same population as non-basketball players, or if they are so different that we have to consider that when looking at height, the population of basketball players should be considered different from the non-basketball player population.

- I know it sounds a little crazy when you first hear about it, but hypothesis testing for one-sample means and proportions is very important. Consider when a policy maker is trying to decide who gets what money. If you can say to him/her that Hispanics are getting significantly less money for education than non-Hispanics, than he/she knows that this population (Hispanics) needs more money than the other population (non-Hispanics), and furthermore that Hispanics should be considered as a different population on this variable than non-Hispanics. In other words, Hispanics do not belong to the same population as non-Hispanics, and should be considered different on this variable.

**8.3    The Five-Step Model for Hypothesis Testing.**

- In order to conduct a one-sample hypothesis test, you need to follow the five-step model for hypothesis testing. These five steps allow you to make some assumptions about your data, sampling distribution, and your sample, as well as to define your null and research hypotheses, and to determine your critical region. The five steps are: (1) making assumptions and meeting test requirements (random sampling, interval-ratio level of measurement, normal sampling distribution), (2) stating the null hypothesis (the statement of "no difference"), (3) selecting the sampling distribution (Z or t) and establishing the critical region, (4) computing the test statistic, and (5) making a decision and interpreting the results of the test ("reject the null hypothesis" or "fail to reject the null hypothesis") (Healey, 2007: 158).

**8.4    One-Tailed and Two-Tailed Tests of Hypothesis.**

- When stating your hypotheses, you need to decide whether you think your sample is merely different from the population or whether you think your sample is higher or lower than the population on the variable in question. With this decision, you set your test to be either one-tailed or two-tailed. In one-tailed tests, you are hypothesizing the direction of the difference: basketball players will be taller than non-basketball players. In two-tailed tests, you are not hypothesizing the direction of the difference, only that a difference exists: basketball players will be different from the population of non-basketball players, but I am unsure whether they will be taller or shorter.

- If the basketball players' mean height is hypothesized to be taller than that of non-basketball players, the difference is on the positive side of the mean. If the basketball players' mean height is hypothesized to be shorter, the difference is on the negative side of the mean.

- Remember: the differences in the populations are compared with Z scores. In Z scores, we are transforming the value of the scores so that the mean is zero and the standard deviation is one. In hypothesis testing, the population mean ($\mu$) is zero, which is compared to our sample mean. In order to do so, the following equation is used:

$$Z = \frac{\overline{X} - \mu}{\dfrac{\sigma}{\sqrt{N}}}$$

- In this equation, the mean of the sample is compared to the mean of the sampling distribution (which is equal to the population mean), and divided by the standard deviation of the sampling distribution. This equation finds the Z score for the sample mean so that we can compare it with the population mean to determine if there is a significant difference between the two.

- Many times, though, the standard deviation of the population ($\Phi$) is not known; therefore, we need to estimate it with the sample standard deviation (s). If the sample size is large (larger than 100) this can be done by substituting s for $\Phi$, and using the following equation (next page):

147

$$Z = \frac{\overline{X} - \mu}{\frac{s}{\sqrt{N-1}}}$$

- Remember: when estimating the standard deviation of the population (Φ) with the standard deviation of the sample (s), we need to correct for the substitution by using N-1 in the denominator rather than just N.

## 8.5 Selecting an Alpha Level.

- Another important point to consider whenever you are using hypothesis tests is the possibility of error. We know that if we set the alpha level to be 0.05, we are expecting to make errors 5% of the time. More precisely, when we reject a null hypothesis or fail to reject a null hypothesis, we know that we are only 95% confident about these conclusions; therefore there is the possibility (albeit a small one) that we are wrong.

- These possibilities are called Type I and Type II errors, or alpha and beta errors, respectively. Type I or alpha errors are made when we reject a null hypothesis that is really true. Or, in English: we conclude that there <u>is</u> a significant difference between our sample and the population when in fact there <u>isn't</u>. Remember: we are only as certain as our alpha level; so we know there's a 5% chance of being wrong--this is <u>that</u> 5%. You have probably heard of "false positives" – the likelihood that a test result will come back positive when in fact it is false -- that is a Type I error.

- Type II or beta errors are made when we fail to reject a null hypothesis that is really false. In English this means that we decide that there is no significant difference between our sample and the population, when in actuality there is.

- Think of it this way: There is only <u>one</u> correct answer in the entire world -- either there is a significant difference between our sample and the population or there is not. We are 95% sure that our results are true, but there is a 5% chance that we have made either an alpha error or a beta error.

- The problem with both alpha and beta errors is that you really never know when you are committing one. Also, if you decrease the likelihood of committing one type of error, you are increasing the likelihood of committing the other type of error. As a general rule, you want to minimize the Type I (alpha) error, rather than the Type II (beta) error. Thus, we set our alpha levels to be low, reducing the likelihood of rejecting a null hypothesis that is actually true.

## 8.6 The Student's *t* Distribution.

- If the sample size is small, though, we need to use what is called Student's t distribution.

- Before we get into the test statistic, how about a fun little story about "Student." Students always ask: Why is it called "Student's t?" and "Who is Student?" As it happens, "Student" was a scientist named William Gossett at the Guinness Brewery in Dublin, Ireland at the turn of the twentieth century who was in charge of testing the yeast culture in the beer. Because he was using very small sample sizes to estimate the yeast, he discovered that he was making many errors in his estimations. Therefore, he created this test statistic and the t distribution curve in order to control for the smaller sample sizes. Because he was still employed by Guinness Brewery, he couldn't publish this new statistic under his name without letting the competing breweries know what he was up to, so he called it "Student's t" (Schutte, 1973[1]).
- All right, back to Student's t. The t-distribution corrects for small sample sizes, because the smaller the sample size, the less representative it is of the population. For example, if you ask 20 people how much money they make, the average of these 20 incomes is not as accurate a measure of income of the entire population as if you had asked 200 people. Therefore, the distribution of the sampling distribution will not be normal in shape, but will instead be flatter. The t-distribution takes into account this flatter shape and bases the test statistic on this distribution.
- The test statistic for Student's t is:

$$t = \frac{\overline{X} - \mu}{\dfrac{s}{\sqrt{N-1}}}$$

- This equation is basically the same as Z. The main difference between Z and t is the critical region, which is smaller for the t distribution than it is for the Z distribution. This again is a product of small sample sizes: we are not convinced that a sample of less than 100 is a true representation of the population, therefore, the smaller t critical region corrects for this problem.

## 8.7 Tests of Hypotheses for Single-Sample Proportions (Large Samples).

- Hypotheses tests of sample means are only appropriate for interval-ratio variables. But not all data are measured at the interval-ratio level. For data measured at the nominal or ordinal level, hypotheses tests of sample proportions are more appropriate.
- Hypotheses tests of proportions are very similar to those for means and use the same five steps. The main difference between the two types of hypotheses tests is that test of proportions always use the Z distribution. A more minor change involves the use of different symbols, because our test is now of a proportion instead of a mean. Because of this change, the formula for the test statistic is also modified, although the rationale behind the formula remains the same (next page):

---

[1] Schutte, Jerald G. 1977. *Everything You Always Wanted to Know about Elementary Statistics (but were afraid to ask)*. Englewood Cliffs: Prentice-Hall, Inc.

$$Z = \frac{P_s - P_u}{\sqrt{\dfrac{P_u(1 - P_u)}{N}}}$$

## MULTIPLE-CHOICE QUESTIONS

1.  The main point in one-sample hypothesis testing is to _____.
    a. determine if there is more than one sample
    b. determine if the sample is taken from the population
    c. determine if one sample is large enough to estimate the population value
    d. determine the sample size necessary to estimate the population

    *Learning Objective: 1*

2.  The Z and t distributions are _____.
    a. exactly the same
    b. different, because Z is flatter than t
    c. different because t is flatter than Z
    d. incomprehensible

    *Learning Objective: 2*

3.  Z (critical) is _____.
    a. the test statistic computed in step four of the hypothesis process
    b. the score that marks the mean
    c. the test that determines the sampling distribution
    d. the score that marks the beginning of the critical region

    *Learning Objective: 2*

4.  Z (obtained) is _____.
    a. the test statistic computed in step four of the hypothesis process
    b. the score that marks the mean
    c. the test that determines the sampling distribution
    d. the score that marks the beginning of the critical region

    *Learning Objective: 2*

5.  The critical region is _____.
    a.  the area between the mean and the obtained value
    b.  the area between the mean and the critical value
    c.  the area beyond the critical value
    d.  the area given in Appendix B

    *Learning Objective: 2*


6.  If you hypothesized that the sample mean was greater than the population mean, you would need to use _____.
    a.  one-sample proportion test
    b.  two-tailed test
    c.  one-tailed test
    d.  population estimators

    *Learning Objective: 4*


7.  The statement in which you say there is "no difference" between the sample and the population is called the _____.
    a.  null hypothesis
    b.  research hypothesis
    c.  assumptions statement
    d.  critical region

    *Learning Objective: 2*


8.  The statement in which you say there is a difference between the sample and the population is called the _____.
    a.  null hypothesis
    b.  research hypothesis
    c.  assumptions statement
    d.  critical region

    *Learning Objective: 2*

9. If you state that the sample mean is hypothesized to be less than the population mean, then the critical region should be _____.
   a. on both sides of the mean
   b. on the right side of the mean
   c. on the left side of the mean
   d. in the center of the distribution

*Learning Objective: 4 & 6*


10. If you have a sample size larger than 100, and a known standard deviation of the population (Φ), you would use _____ to test for significance.
    a. Z distribution for one-sample proportions
    b. t distribution for one-sample proportions
    c. Z distribution for one-sample means
    d. t distribution for one-sample means

*Learning Objective: 5*


11. If you have a sample size smaller than 100, and an unknown standard deviation of the population (Φ), you would use _____ to test for significance.
    a. Z distribution for one-sample proportions
    b. t distribution for one-sample proportions
    c. Z distribution for one-sample means
    d. t distribution for one-sample means

*Learning Objective: 5*


12. If you want to test for significance for nominal data, you would use _____.
    a. Z distribution for one-sample proportions
    b. t distribution for one-sample proportions
    c. Z distribution for one-sample means
    d. t distribution for one-sample means

*Learning Objective: 5*

13. When testing for significance between a sample and a population for nominal data, the term $P_u$ is the _____.
    a. sample proportion
    b. population proportion
    c. sample mean
    d. population mean

    *Learning Objective: 5*

14. The probability of failing to reject a null hypothesis that is, in fact, false, is a _____.
    a. type I, or alpha error
    b. type II, or beta error
    c. critical value
    d. critical region

    *Learning Objective: 7*

15. The probability of rejecting a null hypothesis that is, in fact, true is a _____.
    a. type I, or alpha error
    b. type II, or beta error
    c. critical value
    d. critical region

    *Learning Objective: 7*

16. To say that a result is "significant" means that _____.
    a. the results are due to random chance
    b. the sample is not different from the population
    c. the results are not due to random chance
    d. everyone in the sample is the same

    *Learning Objective: 1*

17. To find the Z score comparing a sample mean to the hypothesized population mean, use the formula _____.

a. $\dfrac{\overline{X} - \mu}{\dfrac{\sigma}{\sqrt{N}}}$

b. $\dfrac{X - \overline{X}}{s}$

c. $\dfrac{\overline{X} - \mu}{s}$

d. $\dfrac{\mu - \overline{X}}{\dfrac{\sigma}{\sqrt{N}}}$

*Learning Objective: 5*

18. For a hypothesis test of a single sample mean, which of the below is not one of the assumptions.
a. random sampling
b. interval-ratio level of measurement
c. normally distributed sampling distribution
d. normally distributed sample statistic

*Learning Objective: 5*

19. When Z(obtained) falls in the critical region, we _____.
a. reject the alternative hypothesis
b. reject the null hypothesis
c. fail to reject the alternative hypothesis
d. fail to reject the null hypothesis

*Learning Objective: 5*

20. When Z(obtained) does not fall in the critical region, we _____.
a. reject the alternative hypothesis
b. reject the null hypothesis
c. fail to reject the alternative hypothesis
d. fail to reject the null hypothesis

*Learning Objective: 5*

## WORK PROBLEMS

Solve each of these problems, using the five-step hypothesis model whenever appropriate. Unless otherwise specified, use an alpha level of 0.05. Assume that all assumptions and test requirements have been met.

1.  A researcher was interested in the GPAs of athletes. He found that in his sample of 350 varsity athletes, the mean GPA was 2.9. He knows the overall mean GPA is 3.1, with a standard deviation of 2.3. Is there a significant difference between athletes and the rest of the population?

    *Learning Objective: 5*

2.  The mean years of education for a sample of 50 CEOs of the top corporations is 17.5 years, with a standard deviation of 1.2 years. The overall mean years of education is 13.4 years. Is the difference significant?

    *Learning Objective: 5*

3.  The proportion of Hispanic students in a survey of 143 students at a college campus in south Texas was 0.35. The proportion of Hispanic students in all universities is 0.09. Does this college have significantly more Hispanics than the rest of the college campuses?

    *Learning Objective: 5*

4.  The average income in a rural community was found to be $30,900 (s=$5,940; N=250). The average income in the population is $30,765. Is this community significantly different from the rest of the population?

    *Learning Objective: 5*

5.  In a study of attitudes toward women working outside the home, a researcher found that 357 men were in favor, while 298 men opposed. In the population, the proportion of people who favor women working outside the home is 0.47. Can we say that the men are significantly different?

    *Learning Objective: 5*

6.  In a recent drug test for a new allergy medicine, the researchers found that of those who were given the medicine (1500 patients), only 6.2% of them reported drowsiness. The drowsiness rate for all other allergy medicines is 10.5%. Because this research is concerned with medication, the researchers wanted to be 99% confident of their findings. Is this medicine significantly different?

    *Learning Objective: 5*

7.  A recent study found that there was no significant difference between a sample of prostitutes and the rest of the population for unwanted pregnancies. If there were 27 prostitutes in the sample, what would be the critical value that would be needed for significance to be found? [Hint: the researchers hypothesized that prostitutes would have a higher rate of unwanted pregnancies than the rest of the population.]

    *Learning Objective: 5*

8.  In a study of 57 homes surrounding a community park, a researcher found that, as he hypothesized, the residents had a lower tolerance for noise than the rest of the city. The t(obtained) for the difference was -2.55. What can you conclude?

    *Learning Objective: 5*

9.  Refer back to the previous problem. How would his findings change if he set the alpha level to be 0.01? How about 0.10?

    *Learning Objective: 5*

10. A group of researchers were interested in modern contraceptive use in a rural village in the mountains of Bolivia, and found that of their sample of 174 women, 34 of the women had knowledge about modern contraceptives. This number was smaller than the estimated 70% of the rest of Bolivian women who had knowledge of modern contraceptives. Is the village from the same population?

    *Learning Objective: 5*

11. In a study of farming practices, it was found that those farmers who used hybrid corn seed had a mean yield of 50 bushels (N=250 and s=10.86). All other farmers had an average yield of 47 bushels. Given an alpha level of 0.01, is there a significant difference between the hybrid seed farmers and the rest of the population?

    *Learning Objective: 5*

12. A researcher hypothesized that people with AIDS would have a lower quality of life than people without AIDS. She collected data using a Quality of Life scale and found that the mean score of the 216 AIDS patients was 6.7, with a standard deviation of 4.3. She knew that the rest of the population had a mean score of 7.1. Do AIDS patients have significantly lower quality of life than the rest of the population? [Assume Quality of Life to be interval-ratio]

*Learning Objective: 5*

13. The mean number of close friends a sample of 47 college women reported as having was 4.9, with a standard deviation of 1.4. The national average of close friends women report having is 3.2. Do women in college report having significantly more close friends than the rest of the population?

*Learning Objective: 5*

14. Refer to the previous problem. How would the results change if you tested whether women in college report having a significantly different number of close friends than the rest of the population?

*Learning Objective: 5*

15. The number of dates the men's dorm of Wesaloosers was pathetically low. Boredom got the best of Tom one long, lonely Friday night, so he decided to take a poll of 30 of his fellow dorm-mates regarding the number of dates each had had in the last month. The mean number of dates was 2.1 (s=0.4). He knew from a report recently published by the university that the mean number of dates for all other dorm-dwellers was 4.6 per month. Are the men of Wesaloosers significantly losing out on dates?

*Learning Objective: 5*

16. Lonely Tom, from Question 13, consoled himself, and other Wesaloosers dorm-mates, by comparing their GPAs with the overall campus GPA. The Wesaloosers dorm sample of the same 30 men had a mean GPA of 3.6 (s=0.7). The mean GPA for the entire campus was 3.1. Can poor Tom take solace in the fact that their dorm grades are higher?

*Learning Objective: 5*

17. Tom still can't figure out why his dorm has the fewest dates, so he then figures out that his dorm of 340 men spend 4 nights per week in the library (4 out of 7). The rest of the campus spends 2.5 nights per week (2.5 out of 7) in the library. Are they spending more time in the library than other students?

*Learning Objective: 5*

18. In 2004, 0.72 (72 percent) of people in a sample of 1706 adults age 45 and older approved of medical marijuana. In that same sample, 0.58 (58 percent) of people admitted to having smoked marijuana. Is approval of medical marijuana due to previous exposure or use? Test to see if the number of those approving medical marijuana significantly differs from the number of those having used marijuana.

*Learning Objective: 5*

## SPSS or MicroCase WORK PROBLEMS

The following work problems can be completed using either SPSS or MicroCase. The solutions will be the same regardless of which one you use. Your instructor may have a preference, so be sure to check with him/her prior to beginning these problems.

1. The U.S. population watches an average of 4 hours of television per day. Using SPSS or MicroCase, determine if the GSS 2006 sample is significantly different from the U.S. population. What if the population mean was only 3 hours; would our sample be significantly different from a population with a mean of 3?

   In order to run a one-sample hypothesis test of a sample mean in SPSS, select the "Analyze" menu and then choose the "One-Sample T Test" option. Select the variable of interest (*tvhours*). Type the population value into the "Test Value:" box. Click the "OK" button.

2. Compare your age to that of the U.S. population mean (*age*). How do you compare to other Americans?

3. Suppose that you read that the mean income (*realinc*) of statistics students is $40,000. How do they compare to other Americans?

4. Compare the occupational prestige (*sei*) of social scientists (65) to that of the U.S. population. How do they compare?

5. Suppose that the average number of children (*childs*) for the U.S. population is 2.2 children. Determine if the GSS is significantly different from the U.S. population.

## MULTIPLE-CHOICE QUESTIONS

| | | | | | | |
|---|---|---|---|---|---|---|
| 1. | b | (p. 178) | 11. | d | (p. 192-193) |
| 2. | c | (p. 193) | 12. | a | (p. 197) |
| 3. | d | (p. 184) | 13. | b | (p. 198) |
| 4. | a | (p. 185) | 14. | b | (p. 191) |
| 5. | c | (p. 184) | 15. | a | (p. 192) |
| 6. | c | (p. 186-187) | 16. | c | (p. 183) |
| 7. | a | (p. 183-184) | 17. | a | (p. 190) |
| 8. | b | (p. 184) | 18. | d | (p. 183) |
| 9. | c | (p. 187-188) | 19. | b | (p. 186) |
| 10. | c | (p. 192-193) | 20. | d | (p. 186) |

## WORK PROBLEMS

1.  We are conducting a two-tail hypothesis test ($\alpha$=0.05) of a sample mean (large samples). Using the five step model for hypothesis testing:

    Step 1: Make assumptions and meet test requirements.
    We were instructed to assume that all test assumptions/requirements (random sampling, interval-ratio level of measurement, normal sampling distribution) have been met. Our variable of interest is GPA, which is an interval-ratio level variable, and because our sample size is greater than 100, we can assume that our sampling distribution is normal.

    Step 2: State the null (and research) hypotheses.
    $H_o$: $\mu = 3.1$
    ($H_1$: $\mu \neq 3.1$)

    Step 3: Select the sampling distribution and establish the critical region.
    Because we are conducting a hypothesis test of a mean (large samples), we will use the Z sampling distribution. For $\alpha$=0.05 (two-tail), Z(critical) equals $\pm 1.96$.

    Step 4: Compute the test statistic.
    $$Z(obtained) = \frac{\overline{X} - \mu}{\frac{s}{\sqrt{N-1}}} = \frac{2.9 - 3.1}{\frac{2.3}{\sqrt{350-1}}} = -1.62$$

    Step 5: Make a decision and interpret the results of the test.
    Because –1.62 does <u>not</u> fall in the critical region, we <u>fail to reject the null hypothesis</u> and conclude that there is no evidence of a significant difference between athletes and non-athletes with regard to GPA.

2.	We are conducting a two-tail hypothesis test ($\alpha=0.05$) of a sample mean (small samples). Using the five step model for hypothesis testing:

Step 1: Make assumptions and meet test requirements.
   We were instructed to assume that all test assumptions/requirements (random sampling, interval-ratio level of measurement, normal sampling distribution) have been met. Our variable of interest is educational attainment (in years), which is an interval-ratio level variable.

Step 2: State the null (and research) hypotheses.
   $H_o$: $\mu = 13.4$
   ($H_1$: $\mu \neq 13.4$)

Step 3: Select the sampling distribution and establish the critical region.
   Because we are conducting a hypothesis test of a mean (small samples), we will use the Student's t sampling distribution. In order to find t(critical) for a two-tailed test at $\alpha=0.05$, we need to find the degrees of freedom (df), which is equal to N-1 or 50-1=49. Looking down the df column in Appendix B, we find that the numbers skip from 40 to 60. In cases such as these, choose the <u>larger</u> of the two critical values (in order to decrease the likelihood of committing a Type I error). For df=40, t(critical)=±2.021, and for df=60, t(critical)=±2.000. Therefore, we should go with the larger critical value of ±2.021.

Step 4: Compute the test statistic.
$$t(obtained) = \frac{\overline{X} - \mu}{\dfrac{s}{\sqrt{N-1}}} = \frac{17.5 - 13.4}{\dfrac{1.2}{\sqrt{50-1}}} = +23.92$$

Step 5: Make a decision and interpret the results of the test.
   Because +23.92 <u>does</u> fall in the critical region, we <u>reject the null hypothesis</u> and conclude that there is evidence of a significant difference between the education of CEOs and the rest of the population.


3.	We are conducting a one-tail hypothesis test ($\alpha=0.05$) of a sample proportion (large samples). Using the five step model for hypothesis testing:

Step 1: Make assumptions and meet test requirements.
   We were instructed to assume that all test assumptions/requirements have been met. Our variable of interest is the proportion of university students that are Hispanic.

Step 2: State the null (and research) hypotheses.
   $H_o$: $P_u = 0.09$
   ($H_1$: $P_u > 0.09$)

160

Step 3: Select the sampling distribution and establish the critical region.
Because we are conducting a hypothesis test of a proportion (large samples), we will use the Z sampling distribution. For α=0.05 (one-tail), Z(critical) equals +1.65. Because we hypothesized that this college had significantly more Hispanic students than the rest of the colleges, we are only interested in the positive scores. Therefore, our one-tailed Z(critical) has a positive sign.

Step 4: Compute the test statistic.

$$Z(obtained) = \frac{P_s - P_u}{\sqrt{\dfrac{(P_u)(1 - P_u)}{N}}} = \frac{0.35 - 0.09}{\sqrt{\dfrac{(0.09)(1 - 0.09)}{143}}} = +10.86$$

Step 5: Make a decision and interpret the results of the test.
Because +10.86 <u>does</u> fall in the critical region, we <u>reject the null hypothesis</u> and conclude that there is evidence of a significant difference between the population at this college (in terms of its proportion of Hispanic students) and other colleges.

4.   We are conducting a two-tail hypothesis test (α=0.05) of a sample mean (large samples). Using the five step model for hypothesis testing:

Step 1: Make assumptions and meet test requirements.
We were instructed to assume that all test assumptions/requirements have been met. Our variable of interest is income, which is an interval-ratio level variable, and because our sample size is greater than 100, we can assume that our sampling distribution is normal.

Step 2: State the null (and research) hypotheses.
  H₀:  $\mu = 30765$
  (H₁:  $\mu \neq 30765$)

Step 3: Select the sampling distribution and establish the critical region.
Because we are conducting a hypothesis test of a mean (large samples), we will use the Z sampling distribution. For α=0.05 (two-tail), Z(critical) equals ±1.96.

Step 4: Compute the test statistic.

$$Z(obtained) = \frac{\overline{X} - \mu}{\dfrac{s}{\sqrt{N-1}}} = \frac{30900 - 30765}{\dfrac{5940}{\sqrt{250-1}}} = +0.36$$

Step 5: Make a decision and interpret the results of the test.
Because +0.36 does <u>not</u> fall in the critical region, we <u>fail to reject the null hypothesis</u> and conclude that there is no evidence of a significant difference between this community and the rest of the population.

5. We are conducting a two-tail hypothesis test ($\alpha$=0.05) of a sample proportion (large samples). Using the five step model for hypothesis testing:

Step 1: Make assumptions and meet test requirements.
We were instructed to assume that all test assumptions/requirements have been met. Our variable of interest is attitudes toward women working outside the home, which can be expressed as a proportion.

Step 2: State the null (and research) hypotheses.
$H_o$: $P_u = 0.47$
($H_1$: $P_u \neq 0.47$)

Step 3: Select the sampling distribution and establish the critical region.
Because we are conducting a hypothesis test of a proportion (large samples), we will use the Z sampling distribution. For $\alpha$=0.05 (two-tail), Z(critical) equals +1.96.

Step 4: Compute the test statistic.

$$Z(obtained) = \frac{P_s - P_u}{\sqrt{\frac{(P_u)(1 - P_u)}{N}}} = \frac{0.55 - 0.47}{\sqrt{\frac{(0.47)(1 - 0.47)}{655}}} = +4.10$$

Step 5: Make a decision and interpret the results of the test.
Because +4.10 <u>does</u> fall in the critical region, we <u>reject the null hypothesis</u> and conclude that there is evidence of a significant difference between men and the rest of the population on attitudes towards women working outside the home.


6. We are conducting a two-tail hypothesis test ($\alpha$=0.01) of a sample proportion (large samples). Using the five step model for hypothesis testing:

Step 1: Make assumptions and meet test requirements.
We were instructed to assume that all test assumptions/requirements have been met. Our variable of interest is the incidence of drowsiness (proportion of the group reporting drowsiness).

Step 2: State the null (and research) hypotheses.
$H_o$: $P_u = 0.105$
($H_1$: $P_u \neq 0.105$)

Step 3: Select the sampling distribution and establish the critical region.
Because we are conducting a hypothesis test of a proportion (large samples), we will use the Z sampling distribution. For $\alpha$=0.01 (two-tail), Z(critical) equals \ +2.58.

Step 4: Compute the test statistic.

$$Z(obtained) = \frac{P_s - P_u}{\sqrt{\dfrac{(P_u)(1-P_u)}{N}}} = \frac{0.062 - 0.105}{\sqrt{\dfrac{(0.105)(1-0.105)}{1500}}} = -5.43$$

Step 5: Make a decision and interpret the results of the test.

Because –5.43 <u>does</u> fall in the critical region, we <u>reject the null hypothesis</u> and conclude that there is evidence of a significant difference between this allergy medication and other allergy medications.

7. With a sample of only 27 prostitutes, we need to use the t distribution. The degrees of freedom value is N-1 or 26 (27-1). At an alpha of 0.05 for a one-tailed test, the t(critical) is +1.706. Therefore, the t(obtained) would have to have been larger than +1.706 for there to be evidence of a significance difference.

8. We are conducting a one-tail hypothesis test (α=0.05) of a sample mean (small samples). Using the five step model for hypothesis testing:

Step 1: Make assumptions and meet test requirements.

We were instructed to assume that all test assumptions/requirements have been met. Our variable of interest is noise tolerance, which is an interval-ratio level variable.

Step 2: State the null (and research) hypotheses.

$H_0$: μ = (unstated in question)

($H_1$: μ < unstated in question)

Step 3: Select the sampling distribution and establish the critical region.

Because we are conducting a hypothesis test of a mean (small samples), we will use the Student's t sampling distribution. In order to find t(critical) for a one-tailed test at α=0.05, we need to find the degrees of freedom (df), which is equal to N-1 or 57-1=56. Looking down the df column in Appendix B, we find that the numbers skip from 40 to 60. In cases such as these, choose the <u>larger</u> of the two critical values. For <u>df=40</u>, t(critical)=-1.684, and for <u>df=60</u>, t(critical)=-1.671. Therefore, we should go with the larger critical value of –1.684. Because the researcher hypothesized that these residents had a lower tolerance for noise than the rest of the city, she/he is only interested in the negative scores. Therefore, our one-tailed t(critical) has a negative sign.

Step 4: Compute the test statistic.

The t(obtained) was given to be –2.55.

Step 5: Make a decision and interpret the results of the test.
Because –2.55 <u>does</u> fall in the critical region, we <u>reject the null hypothesis</u> and conclude that there is evidence of a significant difference between the park residents and other city residents on noise tolerance.

9. But, how do these results change if we change the alpha level to 0.01 and 0.10? If we were to conduct the hypothesis test again, Step 1, Step 2, and Step 4 will not change. Step 3 will change, because our t(critical) will change, and Step 5 will change because we're comparing our t(obtained) to a different t(critical).

For α=0.01:

Step 3: Select the sampling distribution and establish the critical region.
In order to find t(critical) for a one-tailed test at α=0.01, we need to find the degrees of freedom (df), which is equal to N-1 or 57-1=56. For df=40, t(critical)=-2.423, and for df=60, t(critical)=-2.390. Therefore, we should go with the larger critical value of –2.423.

Step 5: Make a decision and interpret the results of the test.
Because –2.55 <u>does</u> fall in the critical region, we <u>reject the null hypothesis</u> and conclude that there is evidence of a significant difference between the park residents and other city residents on noise tolerance.

For α=0.10:

Step 3: Select the sampling distribution and establish the critical region.
In order to find t(critical) for a one-tailed test at α=0.10, we need to find the degrees of freedom (df), which is equal to N-1 or 57-1=56. For df=40, t(critical)=-1.303, and for df=60, t(critical)=-1.296. Therefore, we should go with the larger critical value of –1.303.

Step 5: Make a decision and interpret the results of the test.
Because –2.55 <u>does</u> fall in the critical region, we <u>reject the null hypothesis</u> and conclude that there is evidence of a significant difference between the park residents and other city residents on noise tolerance.

In sum, because the findings were significant at an α of 0.05, we knew that it would also be significant at an α of 0.10. However, we also found that the findings were also significant at an α of 0.01.

10. We are conducting a two-tail hypothesis test (α=0.05) of a sample proportion (large samples). Using the five step model for hypothesis testing:

Step 1: Make assumptions and meet test requirements.
We were instructed to assume that all test assumptions/requirements have been met. Our variable of interest is the proportion of women with knowledge of modern contraceptives. We're told that 34 of the 174 women in this village have knowledge of modern contraceptives; this equals approximately 20% or 0.20.

Step 2: State the null (and research) hypotheses.
$H_o$: $P_u = 0.70$
($H_1$: $P_u \neq 0.70$)

Step 3: Select the sampling distribution and establish the critical region.
Because we are conducting a hypothesis test of a proportion (large samples), we will use the Z sampling distribution. For α=0.05 (two-tail), Z(critical) equals +1.96.

Step 4: Compute the test statistic.

$$Z(obtained) = \frac{P_s - P_u}{\sqrt{\dfrac{(P_u)(1-P_u)}{N}}} = \frac{0.20 - 0.70}{\sqrt{\dfrac{(0.70)(1-0.70)}{174}}} = -14.39$$

Step 5: Make a decision and interpret the results of the test.
Because −14.39 <u>does</u> fall in the critical region, we <u>reject the null hypothesis</u> and conclude that there is evidence of a significant difference between this village and the rest of Bolivian women. Therefore, the evidence suggests that this village does not come from the same population.


11. We are conducting a two-tail hypothesis test (α=0.01) of a sample mean (large samples). Using the five step model for hypothesis testing:

Step 1: Make assumptions and meet test requirements.
We were instructed to assume that all test assumptions/requirements have been met. Our variable of interest is yield of corn (in bushels), which is an interval-ratio level variable, and because our sample size is greater than 100, we can assume that our sampling distribution is normal.

Step 2: State the null (and research) hypotheses.
$H_o$: $\mu = 47$
($H_1$: $\mu \neq 47$)

Step 3: Select the sampling distribution and establish the critical region.
Because we are conducting a hypothesis test of a mean (large samples), we will use the Z sampling distribution. For α=0.01 (two-tail), Z(critical) equals ±2.58.

Step 4: Compute the test statistic.

$$Z(obtained) = \frac{\overline{X} - \mu}{\dfrac{s}{\sqrt{N-1}}} = \frac{50 - 47}{\dfrac{10.86}{\sqrt{250-1}}} = +4.36$$

Step 5: Make a decision and interpret the results of the test.
Because +4.36 <u>does</u> fall in the critical region, we <u>reject the null hypothesis</u> and conclude that there is evidence of a significant difference between farmers who use hybrid corn and farmers who do not.

12.    We are conducting a one-tail hypothesis test ($\alpha = 0.05$) of a sample mean (large samples). Using the five step model for hypothesis testing:

Step 1: Make assumptions and meet test requirements.
We were instructed to assume that all test assumptions/requirements have been met. Our variable of interest is Quality of Life (scale), which we're told is an interval-ratio level variable, and because our sample size is greater than 100, we can assume that our sampling distribution is normal.

Step 2: State the null (and research) hypotheses.
$H_o$:  $\mu = 7.1$
($H_1$:  $\mu < 7.1$)

Step 3: Select the sampling distribution and establish the critical region.
Because we are conducting a hypothesis test of a mean (large samples), we will use the Z sampling distribution. For $\alpha = 0.05$ (one-tail), Z(critical) equals $-1.65$. Because we hypothesized that AIDS patients will have significantly lower quality of life, we are only interested in the negative scores. Therefore, our one-tailed Z(critical) has a negative sign.

Step 4: Compute the test statistic.

$$Z(obtained) = \frac{\overline{X} - \mu}{\dfrac{s}{\sqrt{N-1}}} = \frac{6.7 - 7.1}{\dfrac{4.3}{\sqrt{216-1}}} = -1.37$$

Step 5: Make a decision and interpret the results of the test.
Because $-1.37$ does <u>not</u> fall in the critical region, we <u>fail to reject the null hypothesis</u> and conclude that there is no evidence that AIDS patients are significantly different from the rest of the population on Quality of Life.

13. We are conducting a one-tail hypothesis test ($\alpha=0.05$) of a sample mean (small samples). Using the five step model for hypothesis testing:

Step 1: Make assumptions and meet test requirements.
We were instructed to assume that all test assumptions/requirements have been met. Our variable of interest is number of close friends, which is an interval-ratio level variable.

Step 2: State the null (and research) hypotheses.
$H_o$: $\mu = 3.2$
($H_1$: $\mu > 3.2$)

Step 3: Select the sampling distribution and establish the critical region.
Because we are conducting a hypothesis test of a mean (small samples), we will use the Student's t sampling distribution. In order to find t(critical) for a one-tailed test at $\alpha=0.05$, we need to find the degrees of freedom (df), which is equal to N-1 or 47-1=46. Looking down the df column in Appendix B, we find that the numbers skip from 40 to 60. In cases such as these, choose the larger of the two critical values. For df=40, t(critical)=+1.684, and for df=60, t(critical)=+1.671. Therefore, we should go with the larger critical value of +1.684. Because we hypothesized that women have more close friends, we are only interested in the positive scores. Therefore, our one-tailed t(critical) has a positive sign.

Step 4: Compute the test statistic.
$$t(obtained) = \frac{\overline{X} - \mu}{\dfrac{s}{\sqrt{N-1}}} = \frac{4.9 - 3.2}{\dfrac{1.4}{\sqrt{47-1}}} = +8.24$$

Step 5: Make a decision and interpret the results of the test.
Because +8.24 does fall in the critical region, we reject the null hypothesis and conclude that there is evidence of a significant difference between college women and the rest of the population on the mean number of friends. Indeed, college women do have significantly more friends than the population.

14. How do the results change if we conduct a two-tail hypothesis test instead? Using the five step model for hypothesis testing:

Step 1: Make assumptions and meet test requirements.
We were instructed to assume that all test assumptions/requirements have been met. Our variable of interest is number of close friends, which is an interval-ratio level variable.

Step 2: State the null (and research) hypotheses.
$H_o$: $\mu = 3.2$
($H_1$: $\mu \neq 3.2$)

Step 3: Select the sampling distribution and establish the critical region.
Because we are conducting a hypothesis test of a mean (small samples), we will use the Student's t sampling distribution. In order to find t(critical) for a two-tailed test at α=0.05, we need to find the degrees of freedom (df), which is equal to N-1 or 47-1=46. Looking down the df column in Appendix B, we find that the numbers skip from 40 to 60. In cases such as these, choose the <u>larger</u> of the two critical values. For df=40, t(critical)=±2.021, and for df=60, t(critical)= ±2.000. Therefore, we should go with the larger critical value of ±2.021.

Step 4: Compute the test statistic.

$$t(obtained) = \frac{\overline{X} - \mu}{\frac{s}{\sqrt{N-1}}} = \frac{4.9 - 3.2}{\frac{1.4}{\sqrt{47-1}}} = +8.24$$

Step 5: Make a decision and interpret the results of the test.
Because +8.24 <u>does</u> fall in the critical region, we <u>reject the null hypothesis</u> and conclude that there is evidence of a significant difference between college women and the rest of the population on the mean number of friends. Therefore, our conclusion did not change that much. Although we could say more previously. Now we can only say that college women's number of friends is significantly different, not significantly higher, as we could say before.

15.     We are conducting a one-tail hypothesis test (α=0.05) of a sample mean (small samples). Using the five step model for hypothesis testing:

Step 1: Make assumptions and meet test requirements.
We were instructed to assume that all test assumptions/requirements have been met. Our variable of interest is number of dates, which is an interval-ratio level variable.

Step 2: State the null (and research) hypotheses.
   $H_o$: $\mu = 4.6$
   ($H_1$: $\mu < 4.6$)

Step 3: Select the sampling distribution and establish the critical region.
Because we are conducting a hypothesis test of a mean (small samples), we will use the Student's t sampling distribution. In order to find t(critical) for a one-tailed test at α=0.05, we need to find the degrees of freedom (df), which is equal to N-1 or 30-1=29. Looking down the df column in Appendix B, we find that t(critical)=-1.699. Because we hypothesized that Wesaloosers have fewer dates (that is, we're trying to see if they're "losing out" on dates), we are only interested in the negative scores. Therefore, our one-tailed t(critical) has a negative sign.

Step 4: Compute the test statistic.
$$t(obtained) = \frac{\overline{X} - \mu}{\frac{s}{\sqrt{N-1}}} = \frac{2.1 - 4.6}{\frac{0.4}{\sqrt{30-1}}} = -33.66$$

Step 5: Make a decision and interpret the results of the test.
Because –33.66 <u>does</u> fall in the critical region, we <u>reject the null hypothesis</u> and conclude that there is evidence of a significant difference between the Wesaloosers and the rest of the population on the number of dates.


16. We are conducting a one-tail hypothesis test ($\alpha$=0.05) of a sample mean (small samples). Using the five step model for hypothesis testing:

Step 1: Make assumptions and meet test requirements.
We were instructed to assume that all test assumptions/requirements have been met. Our variable of interest is GPA, which is an interval-ratio level variable.

Step 2: State the null (and research) hypotheses.
$H_o$: $\mu = 3.1$
($H_1$: $\mu > 3.1$)

Step 3: Select the sampling distribution and establish the critical region.
Because we are conducting a hypothesis test of a mean (small samples), we will use the Student's t sampling distribution. In order to find t(critical) for a one-tailed test at $\alpha$=0.05, we need to find the degrees of freedom (df), which is equal to N-1 or 30-1=29. Looking down the df column in Appendix B, we find that t(critical)=+1.699. Because we hypothesized higher GPAs, our one-tailed t(critical) has a positive sign.

Step 4: Compute the test statistic.
$$t(obtained) = \frac{\overline{X} - \mu}{\frac{s}{\sqrt{N-1}}} = \frac{3.6 - 3.1}{\frac{0.7}{\sqrt{30-1}}} = +3.85$$

Step 5: Make a decision and interpret the results of the test.
Because +3.85 <u>does</u> fall in the critical region, we <u>reject the null hypothesis</u> and conclude that there is evidence of a significant difference between the Wesaloosers and the rest of the population on GPA.


17. We are conducting a one-tail hypothesis test of a sample proportion (large samples). Using the five step model for hypothesis testing:

Step 1: Make assumptions and meet test requirements.
We were instructed to assume that all test assumptions/requirements have been met. Our variable of interest is the proportion of weeknights spent in the library. We're told that the Wesaloosers guys spend 4 out of 7 nights in the library, or 0.57 (4/7) of nights.

Step 2: State the null (and research) hypotheses.
The rest of the campus spends 2.5 out of 7 nights in the library, or 0.36 (2.5/7) of nights.
$H_o$: $P_u = 0.36$
($H_1$: $P_u > 0.36$)

Step 3: Select the sampling distribution and establish the critical region.
Because we are conducting a hypothesis test of a proportion (large samples), we will use the Z sampling distribution. For $\alpha = 0.05$ (one-tail), Z(critical) equals +1.65.

Step 4: Compute the test statistic.

$$Z(obtained) = \frac{P_s - P_u}{\sqrt{\dfrac{(P_u)(1 - P_u)}{N}}} = \frac{0.57 - 0.36}{\sqrt{\dfrac{(0.36)(1 - 0.36)}{30}}} = +2.40$$

Step 5: Make a decision and interpret the results of the test.
Because +2.40 <u>does</u> fall in the critical region, we <u>reject the null hypothesis</u> and conclude that there is evidence of a significant difference between the Wesaloosers and the rest of the campus on nights spent in the library. Wesaloosers spend significantly more time in the library than other students.

18.     We are conducting a two-tail hypothesis test of a sample proportion (large samples). Using the five step model for hypothesis testing:

Step 1: Make assumptions and meet test requirements.
We were instructed to assume that all test assumptions/requirements have been met. Our variable of interest is the proportion of respondents approving of medical marijuana.

Step 2: State the null (and research) hypotheses.
$H_o$: $P_u = 0.58$
($H_1$: $P_u \neq 0.58$)

Step 3: Select the sampling distribution and establish the critical region.
Because we are conducting a hypothesis test of a proportion (large samples), we will use the Z sampling distribution. For $\alpha = 0.05$ (two-tail), Z(critical) equals +1.96.

Step 4: Compute the test statistic.

$$Z(obtained) = \frac{P_s - P_u}{\sqrt{\dfrac{(P_u)(1-P_u)}{N}}} = \frac{0.72 - 0.58}{\sqrt{\dfrac{(0.58)(1-0.58)}{1706}}} = +11.72$$

Step 5: Make a decision and interpret the results of the test.
Because +11.72 <u>does</u> fall in the critical region, we <u>reject the null hypothesis</u> and conclude that there is evidence of a significant difference between the number of people supporting medical marijuana and the number of people who have used marijuana.

## SPSS or MicroCase WORK PROBLEMS

1.  Using SPSS, you would want to compare the sample mean (2.96) to the national average mean (4.0) by requesting a one-sample t-test. Although the sample size is larger than 100, the t test distribution becomes normal at that point, thus a t test is an appropriate test.

**One-Sample Test**

| | | | | | 95% Confidence Interval of the Difference | |
| --- | --- | --- | --- | --- | --- | --- |
| | | | | | Test Value = 4 | |
| | t | df | Sig. (2-tailed) | Mean Difference | Lower | Upper |
| tvhours | -11.466 | 672 | .000 | -1.037 | -1.21 | -.86 |

The mean television watched for the population is 4, which is *significantly* different than the sample mean of 2.96. Note in the output the "Sig." value, which is the significance value. If we want to be 95% certain of our results (or wrong 5% of the time or less), then this value must be less than or equal to 0.05 for the results to be considered "significant" and appropriate for being inferred to the population. In short, whenever the "Sig." value is equal to 0.05 or less, we reject the null hypothesis of no difference, and whenever the "Sig." value is greater than 0.05, we fail to reject the null hypothesis of no difference. (This kind of interpretation will occur often with hypothesis tests run in SPSS.) Therefore, the evidence suggests that our sample is significantly different from the U.S. population on mean hours of television watched a day (assuming a population mean of 4 hours of television watched). But does this result change if our population had a mean of 3 hours watched a day?

**One-Sample Test**

| | | | | | 95% Confidence Interval of the Difference | |
| --- | --- | --- | --- | --- | --- | --- |
| | | | | | Test Value = 3 | |
| | t | df | Sig. (2-tailed) | Mean Difference | Lower | Upper |
| tvhours | -.411 | 672 | .681 | -.037 | -.21 | .14 |

The value in the "Sig." column (0.681) is greater than 0.05, so we fail to reject the null hypothesis of no difference. If the population watched only 3 hours of television a day, then the evidence suggests that our sample is not significantly different from the population in television viewership.

2.  Considering that you're probably a college junior, let's plug in the test value as 21 (for 21 years old).

**One-Sample Test**

| | | | | | 95% Confidence Interval of the Difference | |
|---|---|---|---|---|---|---|
| | | | | | Test Value = 21 | |
| | t | df | Sig. (2-tailed) | Mean Difference | Lower | Upper |
| age | 60.386 | 1492 | .000 | 26.431 | 25.57 | 27.29 |

The value in the "Sig." column is less than 0.05, so we reject the null hypothesis of no difference. If the population average is 47.43 years, then the evidence suggests that your age is significantly different from that for the American population.

3.  The SPSS results appear below.

**One-Sample Test**

| | | | | | 95% Confidence Interval of the Difference | |
|---|---|---|---|---|---|---|
| | | | | | Test Value = 40000 | |
| | t | df | Sig. (2-tailed) | Mean Difference | Lower | Upper |
| realinc | -7.872 | 1281 | .000 | -6950.095 | -8682.20 | -5217.99 |

The value in the "Sig." column is less than 0.05, so we reject the null hypothesis of no difference. If the population average is $33,049.90, then the evidence suggests that the income of statistics students is significantly different from that for the American population.

4.   The SPSS results appear below.

**One-Sample Test**

| | Test Value = 65 | | | | | |
|---|---|---|---|---|---|---|
| | t | df | Sig. (2-tailed) | Mean Difference | 95% Confidence Interval of the Difference | |
| | | | | | Lower | Upper |
| sei | -28.707 | 1425 | .000 | -14.9826 | -16.006 | -13.959 |

The value in the "Sig." column is less than 0.05, so we reject the null hypothesis of no difference.  If the population average is 50.017, then the evidence suggests that the occupational prestige of sociologists is significantly different from that for the American population.

5.   The SPSS results appear below.

**One-Sample Test**

| | Test Value = 2.2 | | | | | |
|---|---|---|---|---|---|---|
| | t | df | Sig. (2-tailed) | Mean Difference | 95% Confidence Interval of the Difference | |
| | | | | | Lower | Upper |
| childs | -6.567 | 1496 | .000 | -.283 | -.37 | -.20 |

The value in the "Sig." column is less than 0.05, so we reject the null hypothesis of no difference.  If the population average is 1.92 children, then the evidence suggests that the GSS is significantly different from the American population.

# CHAPTER NINE
## Hypothesis Testing II: The Two-Sample Case

*Learning Objectives*: *By the end of this chapter, you will be able to:*

1. *Identify and cite examples of situations in which the two-sample test of hypothesis is appropriate.*
2. *Explain the logic of hypothesis testing as applied to the two-sample case.*
3. *Explain what an independent random sample is.*
4. *Perform a test of hypothesis for two-sample means or two-sample proportions following the five-step model and correctly interpret the results.*
5. *List and explain each of the factors (especially sample size) that affect the probability of rejecting the null hypothesis. Explain the differences between statistical significance and importance.*

## KEY TERMS

- ✓ Independent Random Samples (p. 207)
- ✓ Pooled Estimate (p. 208)
- ✓ $\sigma_{p-p}$ (p. 214)
- ✓ $\sigma_{\bar{x}-\bar{x}}$ (p. 208)

## CHAPTER SUMMARY

The logic behind two-sample hypothesis testing is very similar to one-sample testing, but instead of testing for a difference between your sample and the rest of the population, you are testing for a difference between <u>two</u> populations. Specifically, you are using two independent random samples of two different populations (i.e. male, female; high income households, low income households). Thus you are using two independent samples to estimate two distinct populations.

This is a very important distinction. Remember: we are using samples of the two populations, but making our hypotheses about the populations. I know that this sounds funny and strange, but try to think of both groups, whether they be males and females, or Protestants and non-Protestants, as two distinct populations, which we are hypothesizing (in the research hypothesis statement) to be different from one another on some variable.

## CHAPTER OUTLINE

**9.1    Introduction.**

- There are three types of two-sample hypothesis tests:  two-sample means for large samples (Z-distribution); two-sample means for small samples (t-distribution); and two-sample proportions for large samples (Z-distribution).  The main distinction in all of these is whether you have means (interval-ratio data) or proportions (nominal data), and whether your samples are large (combined total greater than 100) or small (combined total less than 100).  If you know the answers to each of these, you can easily find the correct test to use.

- As social scientists, we often hear about wage differences between men and women.  If we were interested in testing to see if this difference really does exist in the population, we could use a sample of the general working population in the U.S. (such as the GSS), divide this sample into males and females, and compare the average incomes of both groups.  These two independent samples (males and females) could be thought of as proxies for the entire populations they represent (population of all males, population of all females).   By using the independent samples Z test, we could determine if the difference found between our samples is likely to occur in the population.  Thus, we are using two samples to compare differences between populations.

**9.2    Hypothesis Testing with Sample Means (Large Samples).**

- If you have means for two samples and the combined sample size is greater than 100, then you would use a two-sample means test for large samples (Z-distribution).

- In the null hypothesis statement for two-sample means test you state that there is no difference between the two populations:

$$H_o : \mu_1 = \mu_2$$

- In the research hypothesis, you state either that the two populations are different from one another:

$$(H_1 : \mu_1 \neq \mu_2)$$

or that one population is either greater than or less than the other population:

$$(H_1 : \mu_1 > \mu_2) \ (H_1 : \mu_1 < \mu_2)$$

- Note that in all of these statements we are stating that the <u>populations</u> -- <u>not the samples</u> -- are either similar or different.

- The five steps for conducting a two-sample hypothesis test are very similar to those for conducting a one-sample hypothesis test (chapter eight). In step 1 you state your assumptions and meet test requirements (independent random samples, interval-ratio level of measurement, and a normal sampling distribution). In step 2 you define the null and research hypotheses (as above). After that, in step 3, you select the sampling distribution and define the critical region. Compute the test statistic in step 4. Finally, Make a decision to either reject the null hypothesis (no evidence of a significant difference between the two populations) or fail to reject the null hypothesis (evidence of a significant difference between the two populations) and then interpret the results in step 5.
- However, even before we can conduct the two-sample means test, we must first calculate the pooled estimate, which estimates the standard error by combining information on the dispersion of the two samples (or populations). The two formulas for the pooled estimate distinguish between large sample tests and small sample tests:

$$\text{(large)}\ \sigma_{\bar{x}1-\bar{x}2} = \sqrt{\frac{s_1^2}{N_1-1} + \frac{s_2^2}{N_2-1}} \quad \text{or} \quad \text{(small)}\ \sigma_{\bar{x}1-\bar{x}2} = \sqrt{\frac{N_1 s_1^2 + N_2 s_2^2}{N_1+N_2-2}} \sqrt{\frac{N_1+N_2}{N_1 N_2}}$$

**9.3    Hypothesis Testing with Sample Means (Small Samples).**
- If you have means for two samples and the combined sample size is less than 100, then you would use a two-sample means test for small samples (t-distribution).
- The five steps for conducting a two-sample hypothesis test for small samples are: In step 1 you state your assumptions and meet test requirements (independent random samples, interval-ratio level of measurement, normal sampling distribution, and equal population variances). In step 2 you define the null and research hypotheses (as above). After that, in step 3, you select the sampling distribution and define the critical region. Compute the test statistic in step 4. Finally, Make a decision to either reject the null hypothesis (no evidence of a significant difference between the two populations) or fail to reject the null hypothesis (evidence of a significant difference between the two populations) and then interpret the results in step 5.
- The degrees of freedom for two-sample hypothesis tests of sample means for small samples is df $= n_1 + n_2 - 2$.

**9.4    Hypothesis Testing with Sample Proportions (Large Samples).**
- If you have nominal data (proportions) than you would use either a two-sample proportions test for large samples (Z-distribution) or for small samples (t-distribution). (Although your textbook does not cover two-sample proportions tests for small samples.)
- Two-sample hypotheses tests for proportions use different formulas than do tests for means. The formula for the pooled proportion is:

$$P_u = \frac{N_1 P_{s_1} + N_2 P_{s_2}}{N_1 + N_2}$$

176

The formula for the standard error is:

$$\sigma_{p_1-p_2} = \sqrt{P_u(1-P_u)}\sqrt{\frac{N_1+N_2}{N_1 N_2}}$$

The formula for the test statistic is:

$$\frac{(P_{s_1} - P_{s2}) - (P_{u_1} - P_{u_2})}{\sigma_{p_1-p_2}}$$

## 9.5    The Limitations of Hypothesis Testing: Significance versus Importance.

- The likelihood of rejecting the null hypothesis is affected by four factors: (1) the size of the observed differences; (2) the alpha level, (3) the use of one- or two-tailed tests, and (4) the size of the sample. The size of the observed difference is the only one of these four factors not controlled by the researcher. The alpha level is how comfortable we are with being wrong – higher alpha levels decrease the likelihood of being wrong (Type I errors) but also decrease the likelihood of rejecting the null hypothesis. The likelihood of rejection also increases if the direction of the relationship is hypothesized (that is, using a one-tail test instead of a two-tail test). Finally, increasing sample size increases the likelihood of rejecting the null hypothesis because the larger the sample size, the more representative it is of the population.

## MULTIPLE-CHOICE QUESTIONS

1.    In the two-sample case of hypothesis testing, we are comparing a _____ to a
_____.
   a.  sample, sample
   b.  sample, population
   c.  population, sample
   d.  population, population

*Learning Objective: 1*

2.    In order to test for significance between the two populations, we assume _____.
   a.  simple random samples
   b.  independent random samples
   c.  sampling distribution to be random
   d.  sampling distribution to be positive

*Learning Objective: 2*

3.     One main differences between one-sample hypothesis testing and two-sample hypothesis testing is that in the latter we are _____.
       a.  comparing means
       b.  comparing samples
       c.  comparing a sample to the population
       d.  comparing separate populations

       *Learning Objective: 2*


4.     In the assumption statement for two-sample means (large samples), we state _____.
       a.  independent random samples
       b.  level of measurement is interval-ratio
       c.  equal population variances
       d.  all of these choices

       *Learning Objective: 4*


5.     In the assumption statement for two-sample means (small samples), we additionally state _____.
       a.  the standard errors are equal
       b.  the samples are normal
       c.  the population variances are equal
       d.  the populations are normal

       *Learning Objective: 4*


6.     In the assumption statement for two-sample proportions, we state _____.
       a.  random sampling
       b.  level of measurement is normal
       c.  sampling distribution is normal
       d.  all of these choices

       *Learning Objective: 4*


7.     The null hypothesis states _____.
       a.  no difference between samples
       b.  no difference between populations
       c.  the two samples are different
       d.  the two populations are different

       *Learning Objective: 4*

8.  The research hypothesis states _____.
    a.  no difference between samples
    b.  no difference between populations
    c.  the two samples are different
    d.  the two populations are different

    *Learning Objective: 4*


9.  The Z-distribution is used for _____ samples, whereas the t-distribution is used for _____ samples.
    a.  nominal, interval-ratio
    b.  interval-ratio, nominal
    c.  small, large
    d.  large, small

    *Learning Objective: 4*


10. The equation for degrees of freedom for a two-sample test of means (small samples) is _____.
    a.  $N - 2$
    b.  $N_1(N_1 - 1)$
    c.  $N_1 - N_2$
    d.  $N_1 + N_2 - 2$

    *Learning Objective: 4*


11. When calculating the Z score for two-sample proportions, $P_u$ is _____.
    a.  calculated first
    b.  given
    c.  never used
    d.  all of these choices

    *Learning Objective: 4*

12. The pooled estimate of the standard error of the sampling distribution of the difference in sample means is calculated as _____.

a. $\sqrt{\dfrac{\sigma_1^2}{N_1 - 1} + \dfrac{\sigma_2^2}{N_2 - 1}}$

b. $\sqrt{\dfrac{\sigma_1^2 + \sigma_2^2}{N_1 + N_2}}$

c. $\sqrt{\dfrac{s_1^2}{N_1 - 1} + \dfrac{s_2^2}{N_2 - 1}}$

d. $\sqrt{\dfrac{s_1^2 + s_2^2}{N_1 + N_2}}$

*Learning Objective: 4*

13. In the equation for two-sample means, the term ($\mu_1 - \mu_2$) reduces to _____.
a. the sample means
b. zero
c. not in this equation
d. used only in small samples

*Learning Objective: 4*

14. If a value was found to be significant, we would conclude _____.
a. we fail to reject the null hypothesis
b. we reject the null hypothesis
c. there is no difference between the two populations
d. there is a significant difference between the two samples

*Learning Objective: 4*

15. If we had hypothesized a one-tail test for a large, two-sample means test at an alpha of 0.05, and the Z(obtained) calculated to be +1.62, we would conclude _____.
a. we fail to reject the null hypothesis
b. we reject the null hypothesis
c. there is a significant difference between the two populations
d. there is a significant difference between the two samples

*Learning Objective: 4*

16. For a two-sample proportions test, we need _____.
    a. sample sizes less than 100
    b. the standard deviation of the sampling distribution
    c. degrees of freedom
    d. large sample sizes

    *Learning Objective: 4*

17. In order to test,the difference between males and females with regard to whether they favor or oppose the death penalty, we would use _____.
    a. two-sample means, large samples
    b. two-sample means, small samples
    c. two-sample proportions, large samples
    d. two-sample proportions, small samples

    *Learning Objective: 1*

18. A researcher hypothesizes that people older than 40 are more likely to vote than people younger than 40. The researcher should conduct a _____ test.
    a. a non-directional two-sample test of means
    b. a directional two-sample test of means
    c. a non-directional two-sample test of proportions
    d. a directional two-sample test of proportions

    *Learning Objective: 1*

19. In a two-tail, two-sample hypothesis test of means, the null hypothesis can be written as _____.
    a. $\mu_1 \neq \mu_2$
    b. $\mu_1 > \mu_2$
    c. $\mu_1 \leq \mu_2$
    d. $\mu_1 = \mu_2$

    *Learning Objective: 4*

20.     When $t$(obtained) is less than $t$(critical), we _____ and conclude that
        _____.
        a.  reject null, the two populations are significantly different
        b.  fail to reject null, the two populations are significantly different
        c.  reject null, the two populations are not significantly different
        d.  fail to reject null, the two populations are not significantly different

        *Learning Objective: 4*

## WORK PROBLEMS

Solve each of these problems, using the five-step hypothesis model whenever appropriate. Unless otherwise specified, use an alpha level of 0.05. Assume that all assumptions and test requirements have been met.

1.      Is there a significant difference in population mean income for men and women at the 0.05 level?

| Men | Women |
|---|---|
| $\overline{X}_1$ = $30,450 | $\overline{X}_2$ = $29,950 |
| $s_1$ = 1,595 | $s_2$ = 1,790 |
| $N_1$ = 32 | $N_2$ = 27 |

        *Learning Objective: 4*

2.      The following data are for the number of religious services attended per year for samples of Protestants and Catholics. Does the Catholic population attend services more often than the Protestant population?

| Protestants | Catholics |
|---|---|
| $\overline{X}_1$ = 9.5 | $\overline{X}_2$ = 11.3 |
| $s_1$ = 3.1 | $s_2$ = 4.7 |
| $N_1$ = 147 | $N_2$ = 176 |

        *Learning Objective: 4*

3. A Sociology professor believes that Sociology majors have different GPAs than Psychology majors. Given the data below, is the difference between the populations significant?

| Sociology | Psychology |
|-----------|------------|
| $\overline{X}_1 = 3.1$ | $\overline{X}_2 = 3.0$ |
| $s_1 = 1.4$ | $s_2 = 1.3$ |
| $N_1 = 250$ | $N_2 = 300$ |

*Learning Objective: 4*

4. Educational achievement was measured for both lower income and higher income individuals. Are the populations significantly different?

| Lower | Higher |
|-------|--------|
| $\overline{X}_1 = 12.5$ | $\overline{X}_2 = 14.3$ |
| $s_1 = 3.4$ | $s_2 = 2.5$ |
| $N_1 = 38$ | $N_2 = 47$ |

*Learning Objective: 4*

5. In a survey of attitudes, 42% of the suburban respondents (N=142) said they were afraid of crime. Compare that to the 89% (N=212) of the city respondents. Are city-dwellers more afraid of crime?

*Learning Objective: 4*

6. What is the critical value for a two-sample (two-tail) test with an alpha level of 0.01 ($N_1=15$, $N_2=21$)?

*Learning Objective: 4*

7. What is the critical value for a two-sample (one-tail) test with an alpha level of 0.001 ($N_1 = 50$, $N_2 = 60$).

*Learning Objective: 4*

8.  A researcher found a t(obtained) for a two-sample means test of +2.798. She was interested in a significance level of 0.01 for the one-tailed test ($H_1$: $\mu_1 > \mu_2$). The sample sizes were: $N_1=11$, $N_2=12$. Is the test statistic significant?

*Learning Objective: 4*

9.  What if the significance level for the previous problem was changed to 0.001. Would your conclusion change?

*Learning Objective: 4*

10. In a study of women's knowledge of modern contraceptives in South America, a researcher has samples from Bolivia and Peru. The data below are the proportion of women in their samples who knew about modern contraceptive methods. Is there a significant difference between the populations?

| Bolivia | Peru |
|---|---|
| $P_1$ = 0.71 | $P_2$ = 0.69 |
| $N_1$ = 1350 | $N_2$ = 1205 |

*Learning Objective: 4*

11. Refer back to the previous problem. Is women's knowledge of modern contraceptives significantly higher in Bolivia?

*Learning Objective: 4*

12. In a study of quality of life, a researcher found a difference in scores by racial groups. Are these differences significant at the 0.01 level?

| White | Black |
|---|---|
| $\overline{X}_1$ = 6.9 | $\overline{X}_2$ = 7.2 |
| $s_1$ = 1.4 | $s_2$ = 1.7 |
| $N_1$ = 132 | $N_2$ = 42 |

*Learning Objective: 4*

13. Of the recent graduating class, the mean starting salaries are given below. Do Physical Science majors make significantly more than Social Science majors?

| Physical Science | | Social Science | |
|---|---|---|---|
| $\overline{X}_1$ | = $30,472 | $\overline{X}_2$ | = $28,800 |
| $s_1$ | = 3,273 | $s_2$ | = 2,960 |
| $N_1$ | = 785 | $N_2$ | = 750 |

*Learning Objective: 4*

14. The administration at a large university was worried about the Greek system: specifically, do members of sororities and fraternities spend less hours studying than non-members? A survey of both groups revealed the following data.

| Greek | | Non-Greek | |
|---|---|---|---|
| $\overline{X}_1$ | = 12.3 | $\overline{X}_2$ | = 14.0 |
| $s_1$ | = 4.1 | $s_2$ | = 0.9 |
| $N_1$ | = 51 | $N_2$ | = 53 |

*Learning Objective: 4*

15. A couple is planning to move to a new city several hundred miles away. They want to move into an area with numerous social activities so they can meet friends quickly. They have looked at several neighborhoods in both the suburbs and the city and have collected data on the number of neighborhood parties in each. Which would you suggest they move to?

| Suburbs | | City | |
|---|---|---|---|
| $\overline{X}_1$ | = 4.2 | $\overline{X}_2$ | = 3.5 |
| $s_1$ | = 1.3 | $s_2$ | = 0.9 |
| $N_1$ | = 14 | $N_2$ | = 12 |

*Learning Objective: 4*

16. The guys in Wesaloosa, whom we met in the last chapter, are still upset by their dating record, but even more so now. The guys in Usaloosa dorm have done a little survey of their own dating habits. The following data were collected. Is there a significant difference in the number of dates?

| Wesaloosa | Usaloosa |
|---|---|
| $\overline{X}_1 = 2.1$ | $\overline{X}_2 = 7.5$ |
| $s_1 = 0.4$ | $s_2 = 4.3$ |
| $N_1 = 30$ | $N_2 = 20$ |

*Learning Objective: 4*

17. Again, lonely Tom takes heart in the fact that their GPAs seem higher than the Usaloosa dorm.. Are they?

| Wesaloosa | Usaloosa |
|---|---|
| $\overline{X}_1 = 3.6$ | $\overline{X}_2 = 2.9$ |
| $s_1 = 0.7$ | $s_2 = 1.3$ |
| $N_1 = 30$ | $N_2 = 20$ |

*Learning Objective: 4*

18. Finally, Tom realizes that maybe they are spending too much time in the library. Granted, their grades are higher, but they sure are missing out on the dates. The Usaloosa dorm guys are spending only one night per week in the library (1/7=.14), compared to the Wesaloosa dorm, which spends 4 nights per week (4/7=.57). Is there a significant difference in the time spent in the library between these two populations?

| Wesaloosa | Usaloosa |
|---|---|
| $P_1 = 0.57$ | $P_2 = 0.14$ |
| $N_1 = 340$ | $N_2 = 200$ |

*Learning Objective: 4*

## SPSS or MicroCase WORK PROBLEMS

The following work problems can be completed using either SPSS or MicroCase. The solutions will be the same regardless of which one you use. Your instructor may have a preference, so be sure to check with him/her prior to beginning these problems.

1.  Using the GSS 2006 dataset, test the following hypothesis (*realinc* and *sex*): *Men and women have different incomes.*

2.  Compare the number of children (*childs*) between those with less than a high school diploma and those who completed some graduate work (*degree*).

3.  Compare income (*realinc*) between those living in New England and those living in the south atlantic (*region*).

4.  Test the following hypothesis (*sei* and *race*): *White and black respondents have different occupational prestige.*

## ANSWER KEY

## MULTIPLE-CHOICE QUESTIONS

| | | | | | | |
|---|---|---|---|---|---|---|
| 1. | d | (p. 207) | 11. | a | (p. 214) |
| 2. | b | (p. 207-208) | 12. | c | (p. 209) |
| 3. | d | (p. 206-207) | 13. | b | (p. 208) |
| 4. | d | (p. 209) | 14. | b | (p. 209) |
| 5. | c | (p. 212) | 15. | a | (p. 210) |
| 6. | c | (p. 215) | 16. | d | (p. 214) |
| 7. | b | (p. 209) | 17. | c | (p. 214) |
| 8. | d | (p. 209) | 18. | d | (p. 214-215) |
| 9. | d | (p. 207, 211) | 19. | d | (p. 209) |
| 10. | d | (p. 212) | 20. | d | (p. 213) |

### WORK PROBLEMS

1. We are conducting a two-tail two-sample hypothesis test ($\alpha$=0.05) of sample means (small samples). Using the five step model for hypothesis testing:

Step 1: Make assumptions and meet test requirements.
> We were instructed to assume that all test assumptions/requirements (independent random sampling, interval-ratio level of measurement, normal sampling distribution, equal population variances) have been met.

Step 2: State the null (and research) hypotheses.
> $H_o$: $\mu_1 = \mu_2$
> ($H_1$: $\mu_1 \neq \mu_2$)

Step 3: Select the sampling distribution and establish the critical region.
> Because we are conducting a two-sample test of means (small samples: $N_1 + N_2 < 100$), we use the Student's t sampling distribution. We must first calculate the associated degrees of freedom (df = $N_1 + N_2 - 2$ = 32+27-2 = 57). For $\alpha$=0.05 (two-tail) at df=57, t(critical) equals $\pm 2.021$.

Step 4: Compute the test statistic. (First, calculate the pooled estimate.)

$$\sigma_{\bar{x}1-\bar{x}2} = \sqrt{\frac{N_1 s_1^2 + N_2 s_2^2}{N_1 + N_2 - 2}} \sqrt{\frac{N_1 + N_2}{N_1 N_2}}$$

$$\sigma_{\bar{x}1-\bar{x}2} = \sqrt{\frac{(32)(1595^2) + (27)(1790^2)}{32 + 27 - 2}} \sqrt{\frac{32 + 27}{32 \times 27}} = 448.52$$

$$t(obtained) = \frac{\bar{X}_1 - \bar{X}_2}{\sigma_{\bar{x}1-\bar{x}2}} = \frac{30450 - 29950}{448.52} = +1.11$$

Step 5: Make a decision and interpret the results of the test.
> Because +1.11 does <u>not</u> fall in the critical region, we <u>fail to reject the null hypothesis</u> and conclude that there is no evidence of a significant difference between men's income and women's income.

2. We are conducting a one-tail two-sample hypothesis test ($\alpha$=0.05) of sample means (large samples). Using the five step model for hypothesis testing:

Step 1: Make assumptions and meet test requirements.
> We were instructed to assume that all test assumptions/requirements (independent random sampling, interval-ratio level of measurement, normal sampling distribution) have been met.

Step 2: State the null (and research) hypotheses.
> $H_o$: $\mu_1 = \mu_2$
> ($H_1$: $\mu_1 < \mu_2$)

Step 3: Select the sampling distribution and establish the critical region.
Because we are conducting a two-sample test of means (large samples), we use the Z sampling distribution. For $\alpha=0.05$ (one-tail), Z(critical) equals $-1.65$. The Z(critical) is negative because we hypothesized that Protestants (group 1) would attend services less often than Catholics (group 2).

Step 4: Compute the test statistic. (First, calculate the pooled estimate.)

$$\sigma_{\bar{x}1-\bar{x}2} = \sqrt{\frac{s_1^2}{N_1-1} + \frac{s_2^2}{N_2-1}} = \sqrt{\frac{(3.1)^2}{147-1} + \frac{(4.7)^2}{176-1}} = 0.44$$

$$Z(obtained) = \frac{\overline{X}_1 - \overline{X}_2}{\sigma_{\bar{x}1-\bar{x}2}} = \frac{9.5-11.3}{0.44} = -4.09$$

Step 5: Make a decision and interpret the results of the test.
Because $-4.09$ <u>does</u> fall in the critical region, we <u>reject the null hypothesis</u> and conclude that there is evidence of a significant difference between Protestants' and Catholics' attendance at religious services.

3. We are conducting a two-tail two-sample hypothesis test ($\alpha=0.05$) of sample means (large samples). Using the five step model for hypothesis testing:

Step 1: Make assumptions and meet test requirements.
We were instructed to assume that all test assumptions/requirements (independent random sampling, interval-ratio level of measurement, normal sampling distribution) have been met.

Step 2: State the null (and research) hypotheses.
$H_o$: $\mu_1 = \mu_2$
($H_1$: $\mu_1 \neq \mu_2$)

Step 3: Select the sampling distribution and establish the critical region.
Because we are conducting a two-sample test of means (large samples), we use the Z sampling distribution. For $\alpha=0.05$ (two-tail), Z(critical) equals $\pm1.96$.

Step 4: Compute the test statistic. (First, calculate the pooled estimate.)

$$\sigma_{\bar{x}1-\bar{x}2} = \sqrt{\frac{s_1^2}{N_1-1} + \frac{s_2^2}{N_2-1}} = \sqrt{\frac{(1.4)^2}{250-1} + \frac{(1.3)^2}{300-1}} = 0.12$$

1.96 + 1.69

0.0078714 + 0.0056521

$$Z(obtained) = \frac{\overline{X}_1 - \overline{X}_2}{\sigma_{\bar{x}1-\bar{x}2}} = \frac{3.1-3.0}{0.12} = +0.83$$

Step 5: Make a decision and interpret the results of the test.
Because $+0.83$ does <u>not</u> fall in the critical region, we <u>fail to reject the null hypothesis</u> and conclude that there is no evidence of a significant difference between the GPAs of Sociology and Psychology majors.

4.  We are conducting a two-tail two-sample hypothesis test (α=0.05) of sample means (small samples). Using the five step model for hypothesis testing:

Step 1: Make assumptions and meet test requirements.
    We were instructed to assume that all test assumptions/requirements (independent random sampling, interval-ratio level of measurement, normal sampling distribution, equal population variances) have been met.

Step 2: State the null (and research) hypotheses.
    $H_o$: $\mu_1 = \mu_2$
    ($H_1$: $\mu_1 \neq \mu_2$)

Step 3: Select the sampling distribution and establish the critical region.
    Because we are conducting a two-sample test of means (small samples: $N_1 + N_2 < 100$), we use the Student's t sampling distribution. We must first calculate the associated degrees of freedom (df = $N_1 + N_2 - 2 = 38 + 47 - 2 = 83$). For α=0.05 (two-tail) at df=83, t(critical) equals ±2.000.

Step 4: Compute the test statistic. (First, calculate the pooled estimate.)

$$\sigma_{\bar{x}1-\bar{x}2} = \sqrt{\frac{N_1 s_1^2 + N_2 s_2^2}{N_1 + N_2 - 2}} \sqrt{\frac{N_1 + N_2}{N_1 N_2}} = \sqrt{\frac{(38)(3.4^2) + (47)(2.5^2)}{38 + 47 - 2}} \sqrt{\frac{38 + 47}{38 \times 47}} = 0.65$$

$$t(obtained) = \frac{\overline{X}_1 - \overline{X}_2}{\sigma_{\bar{x}1-\bar{x}2}} = \frac{12.5 - 14.3}{0.65} = -2.77$$

Step 5: Make a decision and interpret the results of the test.
    Because –2.77 does fall in the critical region, we reject the null hypothesis and conclude that there is evidence of a significant difference in the educational achievement of lower and higher income individuals.


5.  We are conducting a one-tail two-sample hypothesis test (α=0.05) of sample proportions (large samples). Using the five step model for hypothesis testing:

Step 1: Make assumptions and meet test requirements.
    We were instructed to assume that all test assumptions/requirements (independent random sampling, interval-ratio level of measurement, normal sampling distribution) have been met.

Step 2: State the null (and research) hypotheses.
    $H_o$: $P_{u1} = P_{u2}$
    ($H_1$: $P_{u1} \gtrless P_{u2}$)

Step 3: Select the sampling distribution and establish the critical region.
Because we are conducting a two-sample test of proportions (large samples), we use the Z sampling distribution. For α=0.05 (one-tail), Z(critical) equals +1.65. The Z(critical) is positive because we hypothesized that city dwellers (group 1) would be more afraid of crime than suburbanites (group 2).

Step 4: Compute the test statistic. (First, calculate the population proportion, $P_u$; then calculate the standard error.)

$$P_u = \frac{N_1 P_{s1} + N_2 P_{s2}}{N_1 + N_2} = \frac{(212)(0.89) + (142)(0.42)}{212 + 142} = 0.70$$

$$\sigma_{p1-p2} = \sqrt{P_u(1 - P_u)} \sqrt{\frac{N_1 + N_2}{N_1 N_2}} = \sqrt{(0.70)(1 - 0.70)} \sqrt{\frac{212 + 142}{(212)(142)}} = 0.05$$

$$Z(obtained) = \frac{P_{s1} - P_{s2}}{\sigma_{p1-p2}} = \frac{0.89 - 0.42}{0.05} = +9.40$$

Step 5: Make a decision and interpret the results of the test.
Because +9.40 <u>does</u> fall in the critical region, we <u>reject the null hypothesis</u> and conclude that there is evidence of a significant difference in fear of crime between people who live in the city and the suburbs.

6. Because the sum of the two samples is less than 100 (15+21=36), we use the t-distribution. To find the critical value, find the degrees of freedom, which is $N_1+N_2-2$, or 34. In the t-distribution table, go down the df column to 30 and 40. The t(critical) for α=0.01 (two-tail) at df=30 is ±2.750. The t(critical) for α=0.01 (two-tail) at df=40 is ±2.704. Taking the larger of the two values, ±2.750 is the critical value.

7. Because the sum of the two samples is greater than 100 (15+21=36), we use the z-distribution. For α=0.001 (two-tail), Z(critical) equals ±3.29.

8. We are conducting a one-tail two-sample hypothesis test (α=0.01) of sample means (small samples). Using the five step model for hypothesis testing:

Step 1: Make assumptions and meet test requirements.
We were instructed to assume that all test assumptions/requirements (independent random sampling, interval-ratio level of measurement, normal sampling distribution, equal population variances) have been met.

Step 2: State the null (and research) hypotheses.
$H_o$: $\mu_1 = \mu_2$
($H_1$: $\mu_1 > \mu_2$)

Step 3: Select the sampling distribution and establish the critical region.
Because we are conducting a two-sample test of means (small samples: $N_1 + N_2 < 100$), we use the Student's t sampling distribution. We must first calculate the associated degrees of freedom (df = $N_1 + N_2 - 2 = 11+12-2 = 21$). For $\alpha = 0.01$ (one-tail) at df=21, t(critical) equals +2.518.

Step 4: Compute the test statistic.
t(obtained) was given to be +2.798

Step 5: Make a decision and interpret the results of the test.
Because +2.798 <u>does</u> fall in the critical region, we <u>reject the null hypothesis</u> and conclude that there is evidence of a significant difference between the two populations.

9.    Now, we are conducting a one-tail two-sample hypothesis test at a 0.001 level. Using the five step model for hypothesis testing:

Step 1: Make assumptions and meet test requirements.
We were instructed to assume that all test assumptions/requirements (independent random sampling, interval-ratio level of measurement, normal sampling distribution, equal population variances) have been met.

Step 2: State the null (and research) hypotheses.
$H_o$:  $\mu_1 = \mu_2$
($H_1$:  $\mu_1 > \mu_2$)

Step 3: Select the sampling distribution and establish the critical region.
Because we are conducting a two-sample test of means (small samples: $N_1 + N_2 < 100$), we use the Student's t sampling distribution. We must first calculate the associated degrees of freedom (df = $N_1 + N_2 - 2 = 11+12-2 = 21$). For $\alpha = 0.001$ (one-tail) at df=21, t(critical) equals +3.527.

Step 4: Compute the test statistic.
t(obtained) was given to be +2.798

Step 5: Make a decision and interpret the results of the test.
Because +2.798 <u>does not</u> fall in the critical region, we <u>fail to reject the null hypothesis</u> and conclude that there is no evidence of a significant difference between the two populations.

10.    We are conducting a two-tail two-sample hypothesis test ($\alpha = 0.05$) of sample proportions (large samples). Using the five step model for hypothesis testing:

Step 1: Make assumptions and meet test requirements.

We were instructed to assume that all test assumptions/requirements (independent random sampling, interval-ratio level of measurement, normal sampling distribution) have been met.

Step 2: State the null (and research) hypotheses.

$H_o$: $P_{u1} = P_{u2}$

($H_1$: $P_{u1} \neq P_{u2}$)

Step 3: Select the sampling distribution and establish the critical region.

Because we are conducting a two-sample test of proportions (large samples), we use the Z sampling distribution. For $\alpha = 0.05$ (two-tail), Z(critical) equals $\pm 1.96$.

Step 4: Compute the test statistic. (First, calculate the population proportion, $P_u$; then calculate the standard error.)

$$P_u = \frac{N_1 P_{s1} + N_2 P_{s2}}{N_1 + N_2} = \frac{(1350)(0.71) + (1205)(0.69)}{1350 + 1205} = 0.70$$

$$\sigma_{p1-p2} = \sqrt{P_u(1-P_u)}\sqrt{\frac{N_1 + N_2}{N_1 N_2}} = \sqrt{(0.70)(1-0.70)}\sqrt{\frac{1350 + 1205}{(1350)(1205)}} = 0.018$$

$$Z(obtained) = \frac{P_{s1} - P_{s2}}{\sigma_{p1-p2}} = \frac{0.71 - 0.69}{0.018} = +1.11$$

Step 5: Make a decision and interpret the results of the test.

Because +1.11 does <u>not</u> fall in the critical region, we <u>fail to reject the null hypothesis</u> and conclude that there is no evidence of a significant difference in knowledge of modern contraceptives for the Bolivian and Peruvian populations.

11.   Now, we are conducting a one-tail two-sample hypothesis test ($\alpha = 0.05$) of sample proportions (large samples). Using the five step model for hypothesis testing:

Step 1: Make assumptions and meet test requirements.

We were instructed to assume that all test assumptions/requirements (independent random sampling, interval-ratio level of measurement, normal sampling distribution) have been met.

Step 2: State the null (and research) hypotheses.

$H_o$: $P_{u1} = P_{u2}$

($H_1$: $P_{u1} > P_{u2}$)

Step 3: Select the sampling distribution and establish the critical region.

Because we are conducting a two-sample test of proportions (large samples), we use the Z sampling distribution. For $\alpha = 0.05$ (one-tail), Z(critical) equals +1.65.

Step 4: Compute the test statistic. (First, calculate the population proportion, $P_u$; then calculate the standard error.)

$$P_u = \frac{N_1 P_{s1} + N_2 P_{s2}}{N_1 + N_2} = \frac{(1350)(0.71) + (1205)(0.69)}{1350 + 1205} = 0.70$$

$$\sigma_{p1-p2} = \sqrt{P_u(1-P_u)}\sqrt{\frac{N_1 + N_2}{N_1 N_2}} = \sqrt{(0.70)(1-0.70)}\sqrt{\frac{1350 + 1205}{(1350)(1205)}} = 0.018$$

$$Z(obtained) = \frac{P_{s1} - P_{s2}}{\sigma_{p1-p2}} = \frac{0.71 - 0.69}{0.018} = +1.11$$

Step 5: Make a decision and interpret the results of the test.
Because +1.11 does <u>not</u> fall in the critical region, we <u>fail to reject the null hypothesis</u> and conclude that there is no evidence of a significant difference in knowledge of modern contraceptives for the Bolivian and Peruvian populations.

12. We are conducting a two-tail two-sample hypothesis test ($\alpha=0.01$) of sample means (large samples). Using the five step model for hypothesis testing:

Step 1: Make assumptions and meet test requirements.
We were instructed to assume that all test assumptions/requirements (independent random sampling, interval-ratio level of measurement, normal sampling distribution) have been met.

Step 2: State the null (and research) hypotheses.
$H_o$: $\mu_1 = \mu_2$
($H_1$: $\mu_1 \neq \mu_2$)

Step 3: Select the sampling distribution and establish the critical region.
Because we are conducting a two-sample test of means (large samples), we use the Z sampling distribution. For $\alpha=0.01$ (two-tail), Z(critical) equals $\pm 2.58$.

Step 4: Compute the test statistic. (First, calculate the pooled estimate.)

$$\sigma_{\bar{x}1-\bar{x}2} = \sqrt{\frac{s_1^2}{N_1 - 1} + \frac{s_2^2}{N_2 - 1}} = \sqrt{\frac{(1.4)^2}{132 - 1} + \frac{(1.7)^2}{42 - 1}} = 0.29$$

$$Z(obtained) = \frac{\bar{X}_1 - \bar{X}_2}{\sigma_{\bar{x}1-\bar{x}2}} = \frac{6.9 - 7.2}{0.29} = -1.03$$

Step 5: Make a decision and interpret the results of the test.
Because −1.03 does <u>not</u> fall in the critical region, we <u>fail to reject the null hypothesis</u> and conclude that there is no evidence of a significant difference in Quality of Life for blacks and whites.

13. We are conducting a one-tail two-sample hypothesis test ($\alpha=0.05$) of sample means (large samples). Using the five step model for hypothesis testing:

Step 1: Make assumptions and meet test requirements.
We were instructed to assume that all test assumptions/requirements (independent random sampling, interval-ratio level of measurement, normal sampling distribution) have been met.

Step 2: State the null (and research) hypotheses.
$H_o$: $\mu_1 = \mu_2$
($H_1$: $\mu_1 > \mu_2$)

Step 3: Select the sampling distribution and establish the critical region.
Because we are conducting a two-sample test of means (large samples), we use the Z sampling distribution. For $\alpha=0.05$ (one-tail), Z(critical) equals +1.65. The Z(critical) is positive because we hypothesized that Physical Science majors (group 1) would have higher starting salaries than Social Science majors (group 2).

Step 4: Compute the test statistic. (First, calculate the pooled estimate.)

$$\sigma_{\bar{x}1-\bar{x}2} = \sqrt{\frac{s_1^2}{N_1-1} + \frac{s_2^2}{N_2-1}} = \sqrt{\frac{(3273)^2}{785-1} + \frac{(2960)^2}{750-1}} = 159.25$$

$$Z(obtained) = \frac{\overline{X}_1 - \overline{X}_2}{\sigma_{\bar{x}1-\bar{x}2}} = \frac{30472 - 28800}{159.25} = +10.50$$

Step 5: Make a decision and interpret the results of the test.
Because +10.5 <u>does</u> fall in the critical region, we <u>reject the null hypothesis</u> and conclude that there is evidence of a significant difference in starting salary for Physical Science and Social Science majors.


14. We are conducting a one-tail two-sample hypothesis test ($\alpha=0.05$) of sample means (large samples). Using the five step model for hypothesis testing:

Step 1: Make assumptions and meet test requirements.
We were instructed to assume that all test assumptions/requirements (independent random sampling, interval-ratio level of measurement, normal sampling distribution) have been met.

Step 2: State the null (and research) hypotheses.
$H_o$: $\mu_1 = \mu_2$
($H_1$: $\mu_1 < \mu_2$)

Step 3: Select the sampling distribution and establish the critical region.
Because we are conducting a two-sample test of means (large samples), we use the Z sampling distribution. For $\alpha=0.05$ (one-tail), Z(critical) equals -1.65. The Z(critical) is negative because we hypothesized that Greeks (group 1) would study fewer hours than Non-Greeks (group 2).

Step 4: Compute the test statistic. (First, calculate the pooled estimate.)

$$\sigma_{\bar{x}1-\bar{x}2} = \sqrt{\frac{s_1^2}{N_1 - 1} + \frac{s_2^2}{N_2 - 1}} = \sqrt{\frac{(4.1)^2}{51-1} + \frac{(0.9)^2}{53-1}} = 0.59$$

$$Z(obtained) = \frac{\overline{X}_1 - \overline{X}_2}{\sigma_{\bar{x}1-\bar{x}2}} = \frac{12.3 - 14.0}{0.59} = -2.88$$

Step 5: Make a decision and interpret the results of the test.
Because −2.88 <u>does</u> fall in the critical region, we <u>reject the null hypothesis</u> and conclude that there is evidence of a significant difference in hours spent studying for the Greek and Non-Greek populations.

15. We are conducting a two-tail two-sample hypothesis test ($\alpha$=0.05) of sample means (small samples). Using the five step model for hypothesis testing:

Step 1: Make assumptions and meet test requirements.
We were instructed to assume that all test assumptions/requirements (independent random sampling, interval-ratio level of measurement, normal sampling distribution, equal population variances) have been met.

Step 2: State the null (and research) hypotheses.
$H_0$: $\mu_1 = \mu_2$
($H_1$: $\mu_1 \neq \mu_2$)

Step 3: Select the sampling distribution and establish the critical region.
Because we are conducting a two-sample test of means (small samples: $N_1$+ $N_2$<100), we use the Student's t sampling distribution. We must first calculate the associated degrees of freedom (df = $N_1$+ $N_2$-2 = 14+12-2 = 24). For $\alpha$=0.05 (two-tail) at df=24, t(critical) equals ±2.064.

Step 4: Compute the test statistic. (First, calculate the pooled estimate.)

$$\sigma_{\bar{x}1-\bar{x}2} = \sqrt{\frac{N_1 s_1^2 + N_2 s_2^2}{N_1 + N_2 - 2}} \sqrt{\frac{N_1 + N_2}{N_1 N_2}} = \sqrt{\frac{(14)(1.3^2) + (12)(0.9^2)}{14 + 12 - 2}} \sqrt{\frac{14 + 12}{14 \times 12}} = 0.46$$

$$t(obtained) = \frac{\overline{X}_1 - \overline{X}_2}{\sigma_{\bar{x}1-\bar{x}2}} = \frac{4.2 - 3.5}{0.46} = +1.52$$

Step 5: Make a decision and interpret the results of the test.
Because +1.52 does <u>not</u> fall in the critical region, we <u>fail to reject the null hypothesis</u> and conclude that there is no evidence of a significant difference in the number of parties in the suburbs and city.

16. We are conducting a two-tail two-sample hypothesis test ($\alpha=0.05$) of sample means (small samples). Using the five step model for hypothesis testing:

Step 1: Make assumptions and meet test requirements.
We were instructed to assume that all test assumptions/requirements (independent random sampling, interval-ratio level of measurement, normal sampling distribution, equal population variances) have been met.

Step 2: State the null (and research) hypotheses.
$H_o$: $\mu_1 = \mu_2$
($H_1$: $\mu_1 \neq \mu_2$)

Step 3: Select the sampling distribution and establish the critical region.
Because we are conducting a two-sample test of means (small samples: $N_1 + N_2 < 100$), we use the Student's t sampling distribution. We must first calculate the associated degrees of freedom (df = $N_1 + N_2 - 2 = 30 + 20 - 2 = 48$). For $\alpha=0.05$ (two-tail) at df=48, t(critical) equals $\pm 2.021$.

Step 4: Compute the test statistic. (First, calculate the pooled estimate.)

$$\sigma_{\bar{x}1-\bar{x}2} = \sqrt{\frac{N_1 s_1^2 + N_2 s_2^2}{N_1 + N_2 - 2}} \sqrt{\frac{N_1 + N_2}{N_1 N_2}} = \sqrt{\frac{(30)(0.4^2) + (20)(4.3^2)}{30 + 20 - 2}} \sqrt{\frac{30 + 20}{30 \times 20}} = 0.81$$

$$t(obtained) = \frac{\overline{X}_1 - \overline{X}_2}{\sigma_{\bar{x}1-\bar{x}2}} = \frac{2.1 - 7.5}{0.81} = -6.67$$

Step 5: Make a decision and interpret the results of the test.
Because –6.67 <u>does</u> fall in the critical region, we <u>reject the null hypothesis</u> and conclude that there is evidence of a significant difference in the number of dates in the Wesaloosa and Usaloosa populations.

17. We are conducting a one-tail two-sample hypothesis test ($\alpha=0.05$) of sample means (small samples). Using the five step model for hypothesis testing:

Step 1: Make assumptions and meet test requirements.
We were instructed to assume that all test assumptions/requirements (independent random sampling, interval-ratio level of measurement, normal sampling distribution, equal population variances) have been met.

Step 2: State the null (and research) hypotheses.
$H_o$: $\mu_1 = \mu_2$
($H_1$: $\mu_1 > \mu_2$)

Step 3: Select the sampling distribution and establish the critical region.
Because we are conducting a two-sample test of means (small samples: $N_1 + N_2 < 100$), we use the Student's t sampling distribution. We must first calculate the associated degrees of freedom (df = $N_1 + N_2 - 2 = 30 + 20 - 2 = 48$). For $\alpha=0.05$ (one-tail) at df=48, t(critical) equals $+1.684$.

Step 4: Compute the test statistic.  (First, calculate the pooled estimate.)

$$\sigma_{\bar{x}1-\bar{x}2} = \sqrt{\frac{N_1 s_1^2 + N_2 s_2^2}{N_1 + N_2 - 2}} \sqrt{\frac{N_1 + N_2}{N_1 N_2}} = \sqrt{\frac{(30)(0.7^2) + (20)(1.3^2)}{30 + 20 - 2}} \sqrt{\frac{30 + 20}{30 \times 20}} = 0.29$$

$$t(obtained) = \frac{\overline{X}_1 - \overline{X}_2}{\sigma_{\bar{x}1-\bar{x}2}} = \frac{3.6 - 2.9}{0.29} = +2.41$$

Step 5: Make a decision and interpret the results of the test.
Because +2.41 <u>does</u> fall in the critical region, we <u>reject the null hypothesis</u> and conclude that there is evidence of a significant difference in the GPAs in the Wesaloosa and Usaloosa populations.


18.    We are conducting a two-tail two-sample hypothesis test ($\alpha$=0.05) of sample proportions (large samples).  Using the five step model for hypothesis testing:

Step 1: Make assumptions and meet test requirements.
We were instructed to assume that all test assumptions/requirements (independent random sampling, interval-ratio level of measurement, normal sampling distribution) have been met.

Step 2: State the null (and research) hypotheses.
$H_o$:  $P_{u1} = P_{u2}$
($H_1$:  $P_{u1} \neq P_{u2}$)

Step 3: Select the sampling distribution and establish the critical region.
Because we are conducting a two-sample test of proportions (large samples), we use the Z sampling distribution.  For $\alpha$=0.05 (two-tail), Z(critical) equals ±1.96.

Step 4: Compute the test statistic.  (First, calculate the population proportion, $P_u$; then calculate the standard error.)

$$P_u = \frac{N_1 P_{s1} + N_2 P_{s2}}{N_1 + N_2} = \frac{(340)(0.57) + (200)(0.14)}{340 + 200} = 0.41$$

$$\sigma_{p1-p2} = \sqrt{P_u(1-P_u)} \sqrt{\frac{N_1 + N_2}{N_1 N_2}} = \sqrt{(0.41)(1-0.41)} \sqrt{\frac{340 + 200}{(340)(200)}} = 0.044$$

$$Z(obtained) = \frac{P_{s1} - P_{s2}}{\sigma_{p1-p2}} = \frac{0.57 - 0.14}{0.044} = +9.77$$

Step 5: Make a decision and interpret the results of the test.
Because +9.77 <u>does</u> fall in the critical region, we <u>reject the null hypothesis</u> and conclude that there is a evidence of a significant difference in time spent in the library for the Wesaloosa and Usaloosa populations.

## SPSS or MicroCase WORK PROBLEMS

If you had problems with any of these questions, be sure to re-read the appropriate sections at the end of chapter 9 in your textbook.

1.  Using the independent samples t-test, the test variable is income (*realinc*). The grouping variable is *sex* (1=males, 2=females).

**Group Statistics**

|  | sex | N | Mean | Std. Deviation | Std. Error Mean |
|---|---|---|---|---|---|
| realinc | MALE | 590 | 35915.14 | 32958.381 | 1356.875 |
|  | FEMALE | 692 | 30607.01 | 30228.088 | 1149.099 |

**Independent Samples Test**

|  | Levene's Test for Equality of Variances | | t-test for Equality of Means | | | | | | | |
|---|---|---|---|---|---|---|---|---|---|---|
|  |  |  |  |  |  |  |  | | 95% Confidence Interval of the Difference | |
|  | F | Sig. | t | df | Sig. (2-tailed) | Mean Difference | Std. Error Difference | | Lower | Upper |
| Equal variances assumed | 4.093 | .043 | 3.006 | 1280 | .003 | 5308.130 | 1765.901 | | 1843.75 | 8772.51 |
| Equal variances not assumed |  |  | 2.985 | 1207.4 | .003 | 5308.130 | 1778.072 | | 1819.68 | 8796.58 |

The mean income for males is $35,915.14 (s=32,958.381). Females have a mean income of $30,607.01 (s=30,228.088). Because in Step 1 of our five-step model we make the assumption that the population variances are normal, we want to use the statistics in the first row of the output ("Equal variances assumed"). Looking under the heading "t-test for Equality of Means," we see that the significance level is less than 0.05 (sig = 0.003), indicating that the we reject the null hypothesis of no difference. Instead we must conclude that there is evidence of a significant difference in income between men and women.

2.  This second problem is a bit trickier because the grouping variable has more than two categories. By clicking on the "Utilities" menu and selecting "Variables" we can get a listing of values for each category. We're interested in those who did not complete high school (group 0) and those with some graduate work (group 4). Be sure to check the values of the groups prior to beginning the t-test.

**Group Statistics**

|  | degree | N | Mean | Std. Deviation | Std. Error Mean |
|---|---|---|---|---|---|
| childs | LT HIGH SCHOOL | 221 | 2.74 | 2.056 | .138 |
|  | GRADUATE | 133 | 1.29 | 1.277 | .111 |

**Independent Samples Test**

| | Levene's Test for Equality of Variances | | t-test for Equality of Means | | | | | | | |
|---|---|---|---|---|---|---|---|---|---|---|
| | F | Sig. | t | df | Sig. (2-tailed) | Mean Difference | Std. Error Difference | 95% Confidence Interval of the Difference | |
| | | | | | | | | Lower | Upper |
| Equal variances assumed | 24.031 | .000 | 7.357 | 352 | .000 | 1.456 | .198 | 1.067 | 1.846 |
| Equal variances not assumed | | | 8.220 | 351.643 | .000 | 1.456 | .177 | 1.108 | 1.805 |

The test variable then is number of children (*childs*) and the grouping variable is *degree*. The average number of children for those with less than a high school diploma is 2.74 (s=2.056); the average for those with a graduate degree is 1.29 (s=1.277). Because the significance value is less than 0.05, we reject the null hypothesis of no difference. The evidence suggests a significant difference is the number of children between those with less education and those with more education.

3.    We're interested in those living in New England (group 1) and those living in the south atlantic (group 5).

**Group Statistics**

| | region | N | Mean | Std. Deviation | Std. Error Mean |
|---|---|---|---|---|---|
| realinc | New England | 48 | 33519.79 | 30710.765 | 4432.717 |
| | South Atlantic | 276 | 32773.75 | 32327.652 | 1945.896 |

**Independent Samples Test**

| | Levene's Test for Equality of Variances | | t-test for Equality of Means | | | | | | | |
|---|---|---|---|---|---|---|---|---|---|---|
| | F | Sig. | t | df | Sig. (2-tailed) | Mean Difference | Std. Error Difference | 95% Confidence Interval of the Difference | |
| | | | | | | | | Lower | Upper |
| Equal variances assumed | .025 | .874 | .149 | 322 | .822 | 746.045 | 5019.472 | -9129.1 | 10621.1 |
| Equal variances not assumed | | | .154 | 66.438 | .878 | 746.045 | 4841.022 | -8918.2 | 10410.3 |

The test variable is income (*realinc*) and the grouping variable is *region*. The average income for those living in New England is 33,519.79, and the average income for those living in the south atlantic is 32,773.75. Because the significance value is greater than 0.05, we fail to reject the null hypothesis of no difference. The evidence suggests no significant difference is income between those living in the two regions.

4.    The SPSS results appear below.

**Group Statistics**

| | race | N | Mean | Std. Deviation | Std. Error Mean |
|---|---|---|---|---|---|
| sei | White | 1051 | 51.888 | 19.9248 | .6146 |
| | Black | 194 | 46.344 | 18.7707 | 1.3477 |

**Independent Samples Test**

| | Levene's Test for Equality of Variances | | t-test for Equality of Means | | | | | | | |
|---|---|---|---|---|---|---|---|---|---|---|
| | | | | | | | | | 95% Confidence Interval of the Difference | |
| | F | Sig. | t | df | Sig. (2-tailed) | Mean Difference | Std. Error Difference | | Lower | Upper |
| Equal variances assumed | 1.889 | .170 | 3.592 | 1243 | .000 | 5.544 | 1.5433 | | 2.5162 | 8.5718 |
| Equal variances not assumed | | | 3.743 | 279.408 | .000 | 5.544 | 1.4812 | | 2.6283 | 8.4597 |

The test variable is occupational prestige (*sei*) and the grouping variable is *race*. The average prestige is 51.888 for Whites or 46.344 for Blacks. Because the significance value is less than 0.05, we reject the null hypothesis of no difference. The evidence suggests a significant difference is occupational prestige between White and Black respondents.

# CHAPTER TEN
## Hypothesis Testing III: The Analysis of Variance

_Learning Objectives_: By the end of this chapter, you will be able to:

1. Identify and cite examples of situations in which ANOVA is appropriate.
2. Explain the logic of hypothesis testing as applied to ANOVA.
3. Perform the ANOVA test, using the five-step model as a guide, and correctly interpret the results.
4. Define and explain the concepts of population variance, total sum of squares, the sum of squares between, the sum of squares within, and mean square estimates.
5. Explain the difference between the statistical significance and the importance of relationships between variables.

## KEY TERMS

- ✓ Analysis of Variance  (p. 232)
- ✓ ANOVA  (p. 232)
- ✓ F Ratio  (p. 236)
- ✓ Mean Square Estimate  (p. 236)
- ✓ One-Way Analysis of Variance  (p. 242)
- ✓ Sum of Squares Between (SSB)  (p. 234)
- ✓ Sum of Squares Within (SSW)  (p. 234)
- ✓ Total Sum of Squares  (p. 234)

## CHAPTER SUMMARY

Analysis of variance, or ANOVA, is very similar in logic to the two-sample t-test (chapter nine). In the two-sample t-test, if we have two samples, let's say males and females, and some interval-ratio variable which we are comparing them on, let's say income, we find the means for each group and compute the test statistic to find out whether or not these two groups are significantly different from each other. Do men make significantly more income than women, or is there no significant difference? These two independent samples, then, are used as proxies for the populations they represent.

The two-sample test is terrific _if_ you have two samples (groups), but often you will have more than two samples (groups). If so, then you need to use the ANOVA test. Like the two-sample test, the ANOVA test measures the differences between groups to determine if there is a significant difference between these groups.

# CHAPTER OUTLINE

## 10.1 Introduction.

- Let's say we are interested in the number of religious services attended per year for different religious groups. If we were to use the two-sample test, we would have to break everyone into two groups, such as Protestants and non-Protestants, or Catholics and non-Catholics, which is ok, but there is probably a lot of variety (or variance) within the group non-Protestant or non-Catholic (that is, the categories are heterogeneous, not homogeneous, as discussed in chapter one). It would be better to compare multiple groups (such as Protestant, Catholic, Jewish, Other, and None) simultaneously. Because there are now five samples (groups), we can no longer compute the two-sample test, but instead need to use the ANOVA test.

## 10.2 The Logic of the Analysis of Variance.

- The ANOVA test analyzes the amount of variation (variance) within each group (sample) and compares it to the amount of variance between the groups (samples). The within variance is the amount of homogeneity within the category, or how similar or dissimilar each case is to the other cases in the same group. The between variance is the amount of homogeneity across the groups, or how similar or dissimilar the groups are from one another.

- Think about the five religious categories above. For a significant difference between these five religious groups to exist, there should be very little variation within each group (Protestants should be like other Protestants; Catholics like other Catholics; etc.), but quite a bit of variation between the groups (Protestants are different from Catholics, Jews, Others, and Nones; Catholics are different from Protestants, Jews, Others, and Nones; etc.). If the within variation is small relative to the between variation, then we can reject the null hypothesis. The null hypothesis states that all of the populations will be the same on the variable of interest, and it is written as (where k is the number of groups, or populations, being compared):

$$H_o: \mu_1 = \mu_2 = \mu_3 = \mu_4 = \mu_5 = \ldots\ldots = \mu_k$$

- The research hypothesis states that at least <u>one</u> of the populations, or groups, is different. The research hypothesis is written out:

$$(H_1: \text{At least one of the population means is different.})$$

- By stating the research hypothesis this way, we reject the null hypothesis if only one population is significantly different from all of the others. Stating the research hypothesis this way also acknowledges that we are not quite sure which population (group) is different, only that we are fairly certain (95% if the alpha level is 0.05) that <u>at least</u> one population (group) <u>is</u> different. It's important to remember that ANOVA does not distinguish among the populations (groups); by itself we cannot determine that group 1 attends significantly more religious service than groups 2 and 4. All we know is that at least one of these groups is significantly different from the others.

## 10.3 The Computation of ANOVA.

- Technically, chapter ten covers one-way analysis of variance, where we seek to determine the effect of a single grouping variable (like religious denomination) on another interval-ratio variable (like religious service attendance). There are other, more complex forms of ANOVA, like two-way analysis of variance, where we seek to determine the effect of two different variables (like religious denomination and race) on a third variable (like religious service attendance).

- Calculating F(obtained) is a fairly involved process. You first calculate the "total variation" in the scores or the total sum of squares (SST), and then you calculate the "variation between groups" or the sum of squares between (SSB). (All formulas are listed below, in order of use.) Use these two quantities to calculate the "variation within groups" or the sum of squares within (SSW). Next, calculate the mean square estimates, the actual estimates of the population variances, by dividing sum of squares within and between by their respective degrees of freedom (dfw=N-k and dfb=k-1). As a last step, calculate F(obtained) by dividing the means square between (MSB) by the mean square within (MSW). This F ratio compares the amount of variance between (across) groups (how much the groups differ from each other) to the amount of variance within groups (how homogenous the groups are). When the F ratio is less than one, we automatically fail to reject the null hypothesis. The extent to which the F ratio is greater than one, the more likely we are to reject the null hypothesis.

- In order to calculate the F ratio for ANOVA, use the following steps.

1. Calculate sum of squares total: $SST = \sum (X_i - \overline{X})^2$ or $SST = \sum X^2 - N(\overline{X})^2$

2. Calculate sum of squares between: $SSB = \sum N_k (\overline{X}_k - \overline{X})^2$

3. Use SST and SSB to calculate sum of squares within: $SSW = SST - SSB$

4. Calculate mean squares between: $MSB = \dfrac{SSB}{dfb}$ where $dfb = k - 1$

5. Calculate mean squares within: $MSW = \dfrac{SSW}{dfw}$ where $dfw = N - k$

6. Calculate the F ratio: $F = \dfrac{MSB}{MSW}$

## 10.4 A Computational Example.

## 10.5 A Test of Significance for ANOVA.

- At this point, let's review the five step model for an ANOVA hypothesis test. In step 1 we state our assumptions and make certain that we meet all of the requirements for the ANOVA test; these are independent random samples, interval-ratio level of measurement (for the variable of interest), normally distributed populations, and equal population variances. ANOVA has strict requirements, but as long as our sample (group) sizes are relatively equal, ANOVA is tolerant of violations of the assumptions. [In situations where your sample sizes are very different, you'll want to use another type of hypothesis tests.] In step 2 we state the null and research hypotheses (see above). In step 3 we select the sampling distribution; we always use the F distribution for ANOVA tests. We also establish the critical region. Using your alpha level (often 0.05), the degrees of freedom within (N-k), and the degrees of freedom between (k-1), find the F(critical) in Appendix D. In step 4 we compute the test statistic, F(obtained), as described below. Finally, in step 5 we compare the F(critical) to the F(obtained), make a decision to reject or fail to reject the null hypothesis, and interpret the results.

## 10.6 An Additional Example for Computing and Testing the Analysis of Variance.

## 10.7 The Limitations of the Test.

- A limitation of ANOVA is that the dependent variable must be measured at the interval-ratio level and that the independent variable's categories should have an (approximately) equal number of cases. Although this latter limitation is often unmet, ANOVA is fortunately tolerant of minor violations of its assumptions.
- Another limitation is one that applies to all significance tests: Just because a difference has been found to be significant does not mean that it's important.
- A final limitation stems from the null hypothesis. Recall that the null says that all population means are equal. When that null is rejected, we do not necessarily know exactly which population mean is different.

## MULTIPLE-CHOICE QUESTIONS

1.  The ANOVA test is often thought of as an extension of the _____.
    a.  Z test
    b.  t test
    c.  standard deviation
    d.  test of one-sample means

    *Learning Objective: 1*

2.     The ANOVA test is concerned with the difference _____ and _____
       groups.
       a. in size, in quantity
       b. in quality, in quantity
       c. between, within
       d. sample means, population means

       *Learning Objective: 2*


3.     ANOVA is an acronym for _____.
       a.  All Numbers Of Variation Across
       b.  Analysis of Numbers Over Another
       c.  Analysis of Other Variables
       d.  Analysis of Variance

       *Learning Objective: 1*


4.     SST is an acronym for _____.
       a.  Sum of Squares Total
       b.  Summary Statistic Total
       c.  Sum of Statistics Together
       d.  Squared Summation of T-test

       *Learning Objective: 4*


5.     SST is equal to _____ + _____.
       a.  t(obtained), t(critical)
       b.  Z(obtained), Z(critical)
       c.  dfw, dfb
       d.  SSB, SSW

       *Learning Objective: 4*


6.     The test statistic computed in ANOVA is the _____.
       a. t-test
       b. F ratio
       c. A ratio
       d. Z test

       *Learning Objective: 3*

7.  One of the primary differences between the t-test and ANOVA is _____.
    a. ANOVA is used for large samples
    b. ANOVA is concerned with only two samples
    c. ANOVA can be used for more than two samples
    d. there is no difference between the two tests

    *Learning Objective: 1*


8.  In order to find the critical region, you need _____ and _____.
    a. degrees of freedom, sample size
    b. degrees of freedom, number of categories
    c. degrees of freedom within, degrees of freedom between
    d. all of these choices

    *Learning Objective: 3*


9.  ANOVA uses the _____ distribution.
    a. A
    b. F
    c. t
    d. Z

    *Learning Objective: 3*


10. The null hypothesis states _____.
    a. all the sample means are equal
    b. the population variances are equal
    c. the population means are equal
    d. at least one of the population means is different

    *Learning Objective: 3*


11. The research hypothesis states _____.
    a. all of the sample means are different
    b. the population variances are unequal
    c. all of the population means is unequal
    d. at least one of the population means is different

    *Learning Objective: 3*

12. The sum of squares _____ reflects the pattern of variation of scores in the same category.
    a. between
    b. within
    c. total
    d. mean

*Learning Objective: 4*


13. Degrees of freedom within is found with _____ equation.
    a. $N_k - N$
    b. $N_1 + N_2 + N_3 + ...N_n - n$
    c. $N - 1$
    d. $N - k$

*Learning Objective: 3*


14. Degrees of freedom between is found with _____ equation.
    a. $k - 1$
    b. $N_{k1} + N_{k2} + Nk_3 + ...N_{kn} - k$
    c. $N - 1$
    d. $N - k$

*Learning Objective: 3*


15. The test statistic for ANOVA is equal to _____.
    a. dfb/dfw
    b. dfw/dfb
    c. mean square within/mean square between
    d. mean square between/mean square within

*Learning Objective: 3*


16. In symbolic form, the null hypothesis can be written as _____.
    a. $\mu_1 = \mu_2$
    b. $\mu_1 = \mu_2 = \mu_3 = \mu_4 = ... = \mu_k$
    c. $\pi_1 = \pi_2$
    d. $\mu_1 \neq \mu_2$

*Learning Objective: 3*

17. The total sum of squares is calculated with _____.
   a. $\sum X^2 - N\bar{X}^2$
   b. SST – SSB
   c. $\sum N_k (\bar{X}_k - \bar{X})^2$
   d. $\sum (X_i - \bar{X})^2$

   *Learning Objective: 3*

18. The between sum of squares is calculated with _____.
   a. $\sum X^2 - N\bar{X}^2$
   b. SST – SSB
   c. $\sum N_k (\bar{X}_k - \bar{X})^2$
   d. $\sum (X_i - \bar{X})^2$

   *Learning Objective: 3*

19. The within sum of squares is calculated with _____.
   a. $\sum X^2 - N\bar{X}^2$
   b. SST – SSB
   c. $\sum N_k (\bar{X}_k - \bar{X})^2$
   d. $\sum (X_i - \bar{X})^2$

   *Learning Objective: 3*

20. When a sum of squares value is divided by its respective degrees of freedom. The result is called _____.
   a. mean square estimate
   b. total sum of squares
   c. F ratio
   d. ANOVA

   *Learning Objective: 4*

# WORK PROBLEMS

For all work problems, assume that all assumptions and requirements have been met. Unless otherwise indicated, assume that alpha is 0.05.

1.      A researcher collected data on educational attainment (in years) in the U.S. Is there a significant difference between the groups?

| White | Black | Hispanic |
|-------|-------|----------|
| 12 | 12 | 8 |
| 10 | 12 | 5 |
| 14 | 11 | 12 |
| 16 | 14 | 16 |
| 17 | 16 | 11 |
| 12 | 12 | 10 |
| 16 | 13 | 8 |
| 16 | 10 | 14 |
| 15 | 16 | 18 |
| 8 | 17 | 19 |
| 12 | 16 | 11 |

*Learning Objective: 3*

2.   Many counselors suggest taking statistics in either the sophomore or junior year. Below
     are the average grades for one statistics class. Is there a significant difference?

| Freshman | Sophomores | Juniors | Seniors |
|---|---|---|---|
| 79 | 81 | 90 | 87 |
| 76 | 80 | 93 | 91 |
| 82 | 87 | 82 | 69 |
| 91 | 94 | 57 | 82 |
| 90 | 62 | 62 | 86 |
| 77 | 90 | 79 | 88 |
| 78 | 88 | 71 | 71 |
| 56 | 80 | 74 | 75 |
| 84 | 76 | 81 | 62 |
| 87 | 65 | 81 | 54 |
| 72 | 70 | 85 | 47 |
| 29 | 69 | 69 | 81 |
|  | 72 | 97 | 82 |
|  | 82 | 58 | 77 |
|  | 81 | 41 | 71 |
|  | 78 | 91 | 75 |
|  | 77 | 73 | 76 |
|  | 79 | 82 |  |
|  |  | 84 |  |
|  |  | 81 |  |

*Learning Objective: 3*

3.   How do the data in the previous problem pose problems for ANOVA?

*Learning Objective: 2*

4.  At a recent wine festival, three wineries were judged on taste. Below are the scores given to each winery. Is there a significant difference? (Assume the data to be interval-ratio.)

| Winery 1 | Winery 2 | Winery3 |
|----------|----------|---------|
| 10       | 9        | 10      |
| 9        | 9        | 10      |
| 5        | 10       | 7       |
| 8        | 8        | 6       |
| 7        | 7        | 3       |
| 8.5      | 9        | 5.5     |
| 3        | 8.5      | 10      |
| 10       | 6        | 4.5     |
| 9        | 5        | 7       |
| 5        | 7.5      | 8.5     |
| 6        | 7        | 6       |
| 8        | 9        | 4       |
| 8        | 8        | 5       |
| 7.5      | 8        | 9       |
| 6.5      | 7.5      | 3       |

*Learning Objective: 3*

5.  Based on the results for the previous problem, can you identify the winning winery?

*Learning Objective: 3*

6. The GPA of three dorms has been monitored by the administration of a university. Is there cause for worry?

| Dorm 1 | Dorm 2 | Dorm 3 |
| --- | --- | --- |
| 3.5 | 2.0 | 2.0 |
| 4.0 | 1.6 | 4.0 |
| 3.0 | 3.0 | 3.5 |
| 3.0 | 2.0 | 3.5 |
| 2.7 | 3.2 | 3.2 |
| 3.1 | 3.4 | 3.1 |
| 3.2 | 2.1 | 3.0 |
| 3.3 | 2.6 | 1.8 |
| 3.5 | 2.8 | 2.5 |
| 3.2 | 1.8 | 2.8 |
| 4.0 | 2.4 | 3.1 |
| 3.8 | 3.1 | 2.8 |
| 3.6 | 1.0 | 2.6 |
| 2.8 | 0.8 | 2.5 |

*Learning Objective: 3*

7. Based on the results for the previous problem, can you pinpoint a particular dorm to be a test case for a new study program?

*Learning Objective: 3*

8.  Wesaloosa and Usaloosa dorms have been butting heads for quite awhile. Now a third dorm wants to join in. They have collected data on the number of dates per month from a sample of their residents. Is there a significant difference?

| Wesaloosa Dorm | Usaloosa Dorm | Alsaloosa Dorm |
|---|---|---|
| 4 | 4 | 4 |
| 1 | 6 | 1 |
| 0 | 5 | 8 |
| 5 | 8 | 0 |
| 2 | 0 | 0 |
| 1 | 6 | 3 |
| 3 | 2 | 4 |
| 4 | 0 | 2 |
| 1 | 5 | 1 |
| 2 | 1 | 8 |
| 0 | 6 | 6 |
| 4 | 8 | 5 |
| 3 | 4 | 4 |
| 8 | 4 | 1 |

*Learning Objective: 3*

9.  The number of social events per year for three areas of residence have been collected. Is there a significant difference in the number of social events by area of residence?

| Rural | Urban | Suburban |
|---|---|---|
| 0 | 4 | 12 |
| 3 | 6 | 0 |
| 1 | 5 | 1 |
| 0 | 8 | 0 |
| 1 | 1 | 3 |
| 5 | 0 | 4 |
| 12 | 4 | 1 |
|  | 2 | 6 |

*Learning Objective: 3*

214

10. There's been a lot of talk lately about quick weight-loss programs. Is there a significant difference between these groups?

| Weight Losers | Slim Quik | Forget Fat | No Dishin' | Pound No More |
|---|---|---|---|---|
| 5 | 5 | 20 | 2 | 10 |
| 30 | 10 | 21 | 5 | 14 |
| 29 | 8 | 16 | 4 | 12 |
| 26 | 6 | 15 | 7 | 8 |
| 25 | 2 | 12 | 10 | 4 |
| 10 | 10 | 8 | 6 | 7 |
| 8 | 8 | 4 | 7 | 5 |
| 7 | 7 | 31 | 8 | 22 |
| 5 | 4 | 15 | 3 | 16 |
| 22 | 12 | 10 | 2 | 5 |

*Learning Objective: 3*


11. Is there a significant difference in the number of calls placed to the police department by area of residence?

| Inner City | Suburban | Urban | Rural |
|---|---|---|---|
| 4 | 0 | 7 | 0 |
| 0 | 7 | 8 | 0 |
| 8 | 6 | 9 | 5 |
| 12 | 4 | 11 | 1 |
| 24 | 12 | 1 | 3 |
| 52 | 52 | 5 | 2 |
| 100 | 8 | 0 | 4 |
| 14 | 18 | 0 | 6 |
| 87 | 24 | 4 | 0 |
| 30 | 1 | 6 | 1 |
| 2 | 6 | 3 | 3 |
| 1 | 5 | 8 | 4 |
| 18 | 10 | 1 | 6 |

*Learning Objective: 3*


12. Using the information from problem 8, would there still be a significant difference in the number of calls placed to the police department by area of residence if alpha equaled 0.01?

*Learning Objective: 3*


215

13.    Complete the following table.

| dfw | dfb | F(critical) α=0.05 | F(critical) α=0.01 |
|-----|-----|-----|-----|
| 1 | 1 | _____ | _____ |
| 10 | 4 | _____ | _____ |
| 35 | 10 | _____ | _____ |
| 100 | 5 | _____ | _____ |
| 58 | 3 | _____ | _____ |

*Learning Objective: 4*

## SPSS or MicroCase WORK PROBLEMS

The following work problems can be completed using either SPSS or MicroCase. The solutions will be the same regardless of which one you use. Your instructor may have a preference, so be sure to check with him/her prior to beginning these problems.

1.    Using the GSS 2006 dataset, test the hypothesis that there is a significant difference in the number of children one has (*childs*) among those with different education levels (*degree*). Be sure to compare means and standard deviations across the populations.

2.    Is there a significant difference in the hours of television watched per day (*tvhours*) among those with more education (*degree*)? Be sure to compare means and standard deviations across the populations.

3.    Is there a significant difference in the income (*realinc*) among those with more education (*degree*)? Be sure to compare means and standard deviations across the populations.

4.    Is there a significant difference in the occupational prestige (*sei*) among those with more education (*degree*)? Be sure to compare means and standard deviations across the populations.

## ANSWER KEY

### MULTIPLE-CHOICE QUESTIONS

| | | | | | | |
|---|---|---|---|---|---|---|
| 1. | b | (p. 232) | 11. | d | (p. 238) |
| 2. | c | (p. 233-234) | 12. | b | (p. 234) |
| 3. | d | (p. 232) | 13. | d | (p. 235) |
| 4. | a | (p. 234) | 14. | a | (p. 236) |
| 5. | d | (p. 234) | 15. | a | (p. 236) |
| 6. | b | (p. 236) | 16. | b | (p. 238) |
| 7. | d | (p. 232) | 17. | a | (p. 237) |
| 8. | c | (p. 235-236) | 18. | c | (p. 237) |
| 9. | b | (p. 236) | 19. | b | (p. 237) |
| 10. | c | (p. 238) | 20. | a | (p. 236) |

### WORK PROBLEMS

1.

| | White | Black | Hispanic |
|---|---|---|---|
| | 12 | 12 | 8 |
| | 10 | 12 | 5 |
| | 14 | 11 | 12 |
| | 16 | 14 | 16 |
| | 17 | 16 | 11 |
| | 12 | 12 | 10 |
| | 16 | 13 | 8 |
| | 16 | 10 | 14 |
| | 15 | 16 | 18 |
| | 8 | 17 | 19 |
| | 12 | 16 | 11 |
| $\sum X_i =$ | 221 | 124 | 84 |
| $\sum X^2 =$ | 3,361 | 1,594 | 970 |
| $\overline{X}_k =$ | 14.73 | 12.4 | 10.5 |
| $\overline{X} =$ | 13.0 | | |

$$SST = \sum X^2 - N(\overline{X})^2 = (3361 + 1594 + 970) - (33)(13)^2 = 348$$

$$SSB = \sum N_k (\overline{X}_k - \overline{X})^2 = (15)(14.73 - 13)^2 + (10)(12.4 - 13)^2 + (8)(10.5 - 13)^2 = 98.49$$

$$SSW = SST - SSB = 348 - 98.49 = 249.51$$

$$MSB = \frac{SSB}{k-1} = \frac{98.49}{3-1} = 49.25 \qquad MSW = \frac{SSW}{N-k} = \frac{249.51}{33-3} = 8.32$$

$$F = \frac{MSB}{MSW} = \frac{49.25}{8.32} = 5.92$$

Step 1: Make assumptions and meet test requirements.

We were instructed to assume that all test assumptions/requirements (independent random samples, interval-ratio level of measurement, normal population distribution, equal population variances) have been met.

Step 2: State the null (and research) hypotheses.

$H_0$: $\mu_1 = \mu_2 = \mu_3$

($H_1$: at least one of the population means is different)

Step 3: Select the sampling distribution and establish the critical region.

We always use the F distribution for analyses of variance. For $\alpha=0.05$ at dfw=30 and dfb=2, F(critical) equals 3.32.

Step 4: Compute the test statistic (see calculations above).

F(obtained)=5.92

Step 5: Make a decision and interpret the results of the test.

Because 5.92 does fall in the critical region, we reject the null hypothesis. There is evidence that at least one of the population mean educational attainments is significantly different.

2.

| | Freshman | Sophomores | Juniors | Seniors |
|---|---|---|---|---|
| | 79 | 81 | 90 | 87 |
| | 76 | 80 | 93 | 91 |
| | 82 | 87 | 82 | 69 |
| | 91 | 94 | 57 | 82 |
| | 90 | 62 | 62 | 86 |
| | 77 | 90 | 79 | 88 |
| | 78 | 88 | 71 | 71 |
| | 56 | 80 | 74 | 75 |
| | 84 | 76 | 81 | 62 |
| | 87 | 65 | 81 | 54 |
| | 72 | 70 | 85 | 47 |
| | 29 | 69 | 69 | 81 |
| | | 72 | 97 | 82 |
| | | 82 | 58 | 77 |
| | | 81 | 41 | 71 |
| | | 78 | 91 | 75 |
| | | 77 | 73 | 76 |
| | | 79 | 82 | |
| | | | 84 | |
| | | | 81 | |
| $\sum X_i =$ | 90.1 | 1421 | 1531 | 1274 |
| $\sum X^2 =$ | 70921 | 113239 | 120837 | 97766 |
| $\overline{X}_k =$ | 75.1 | 83.6 | 76.6 | 74.9 |
| $\overline{X} =$ | 76.5 | | | |

$$SST = \sum X^2 - N(\overline{X})^2 = (70921 + 113239 + 120837 + 97766) - (67)(76.5)^2 = 10663$$

$$SSB = \sum N_k(\overline{X}_k - \overline{X})^2$$
$$= (12)(75.1 - 76.5)^2 + (18)(83.6 - 76.5)^2 + (20)(76.6 - 76.5)^2 + (17)(74.9 - 76.5)^2$$
$$= 1042.36$$

$$SSW = SST - SSB = 10663 - 1042.36 = 9620.64$$

$$MSB = \frac{SSB}{k-1} = \frac{1042.36}{4-1} = 347.4 \qquad MSW = \frac{SSW}{N-k} = \frac{9620.64}{67-4} = 152.7$$

$$F = \frac{MSB}{MSW} = \frac{347.4}{152.7} = 2.27$$

Step 1: Make assumptions and meet test requirements.
　　　　We were instructed to assume that all test assumptions/requirements (independent random samples, interval-ratio level of measurement, normal population distribution, equal population variances) have been met.

Step 2: State the null (and research) hypotheses.
　　　　$H_o$:  $\mu_1 = \mu_2 = \mu_3 = \mu_4$
　　　　($H_1$:  at least one of the population means is different)

Step 3: Select the sampling distribution and establish the critical region.
　　　　We always use the F distribution for analyses of variance.  For $\alpha=0.05$ at dfw=63 and dfb=3, F(critical) equals 2.76.

Step 4: Compute the test statistic (see calculations above).
　　　　F(obtained)=2.27

Step 5: Make a decision and interpret the results of the test.
　　　　Because 2.27 does <u>not</u> fall in the critical region, we <u>fail to reject the null hypothesis</u>.  There is no evidence of a significant difference among the statistics grades across the school years.

3.　　　The categories of the independent variable (*class standing*) are unequally sized.  While this can be a limitation for ANOVA, ANOVA is fairly immune to such limitations.

4.

| | Winery 1 | Winery 2 | Winery3 |
|---|---|---|---|
| | 10 | 9 | 10 |
| | 9 | 9 | 10 |
| | 5 | 10 | 7 |
| | 8 | 8 | 6 |
| | 7 | 7 | 3 |
| | 8.5 | 9 | 5.5 |
| | 3 | 8.5 | 10 |
| | 10 | 6 | 4.5 |
| | 9 | 5 | 7 |
| | 5 | 7.5 | 8.5 |
| | 6 | 7 | 6 |
| | 8 | 9 | 4 |
| | 8 | 8 | 5 |
| | 7.5 | 8 | 9 |
| | 6.5 | 7.5 | 3 |
| $\sum X_i =$ | 110.5 | 118.5 | 98.5 |
| $\sum X^2 =$ | 832.75 | 959.75 | 732.75 |
| $\overline{X}_k =$ | 7.4 | 7.9 | 6.6 |
| $\overline{X} =$ | 7.28 | | |

$$SST = \sum X^2 - N(\overline{X})^2 = (832.75 + 959.75 + 732.75) - (45)(7.28)^2 = 192.45$$

$$SSB = \sum N_k(\overline{X}_k - \overline{X})^2 = (15)(7.4 - 7.28)^2 + (15)(7.9 - 7.28)^2 + (15)(6.6 - 7.28)^2 = 12.92$$

$$SSW = SST - SSB = 192.45 - 12.92 = 179.53$$

$$MSB = \frac{SSB}{k-1} = \frac{12.92}{3-1} = 6.46 \qquad MSW = \frac{SSW}{N-k} = \frac{179.53}{45-3} = 4.27$$

$$F = \frac{MSB}{MSW} = \frac{6.46}{4.27} = 1.51$$

Step 1: Make assumptions and meet test requirements.
   We were instructed to assume that all test assumptions/requirements (independent random samples, interval-ratio level of measurement, normal population distribution, equal population variances) have been met.

Step 2: State the null (and research) hypotheses.
   $H_o$: $\mu_1 = \mu_2 = \mu_3$
   ($H_1$: at least one of the population means is different)

Step 3: Select the sampling distribution and establish the critical region.
We always use the F distribution for analyses of variance. For α=0.05 at dfw=42 and dfb=2, F(critical) equals 3.23.

Step 4: Compute the test statistic (see calculations above).
F(obtained)=1.51

Step 5: Make a decision and interpret the results of the test.
Because 1.51 does <u>not</u> fall in the critical region, we <u>fail to reject the null hypothesis</u>. There is no evidence of a significant difference among the ratings of these three wineries.

5.      Because we failed to reject the null hypothesis, the population means are all equal, meaning that no single winery did better or worse than the others.

6.

| | Dorm 1 | Dorm 2 | Dorm 3 |
|---|---|---|---|
| | 3.5 | 2.0 | 2.0 |
| | 4.0 | 1.6 | 4.0 |
| | 3.0 | 3.0 | 3.5 |
| | 3.0 | 2.0 | 3.5 |
| | 2.7 | 3.2 | 3.2 |
| | 3.1 | 3.4 | 3.1 |
| | 3.2 | 2.1 | 3.0 |
| | 3.3 | 2.6 | 1.8 |
| | 3.5 | 2.8 | 2.5 |
| | 3.2 | 1.8 | 2.8 |
| | 4.0 | 2.4 | 3.1 |
| | 3.8 | 3.1 | 2.8 |
| | 3.6 | 1.0 | 2.6 |
| | 2.8 | 0.8 | 2.5 |
| $\sum X_i =$ | 46.7 | 31.8 | 40.4 |
| $\sum X^2 =$ | 158.01 | 76.66 | 121.14 |
| $\overline{X}_k =$ | 3.34 | 2.27 | 2.86 |
| $\overline{X} =$ | 2.83 | | |

$$SST = \sum X^2 - N(\overline{X})^2 = (158.01 + 79.66 + 121.14) - (42)(2.83)^2 = 22.39$$

$$SSB = \sum N_k \left( \overline{X}_k - \overline{X} \right)^2 = (14)(3.34 - 2.83)^2 + (14)(2.27 - 2.83)^2 + (14)(2.86 - 72.83)^2$$
$$= 7.86$$

$$SSW = SST - SSB = 22.39 - 7.86 = 14.53$$

$$MSB = \frac{SSB}{k-1} = \frac{7.86}{3-1} = 3.93 \qquad MSW = \frac{SSW}{N-k} = \frac{14.53}{42-3} = 0.37$$

$$F = \frac{MSB}{MSW} = \frac{3.93}{0.37} = 10.62$$

Step 1: Make assumptions and meet test requirements.
   We were instructed to assume that all test assumptions/requirements (independent random samples, interval-ratio level of measurement, normal population distribution, equal population variances) have been met.

Step 2: State the null (and research) hypotheses.
   $H_o$: $\mu_1 = \mu_2 = \mu_3$
   ($H_1$: at least one of the population means is different)

Step 3: Select the sampling distribution and establish the critical region.
   We always use the F distribution for analyses of variance. For $\alpha=0.05$ at dfw=39 and dfb=2, F(critical) equals 3.32.

Step 4: Compute the test statistic (see calculations above).
   F(obtained)=10.62

Step 5: Make a decision and interpret the results of the test.
   Because 10.62 <u>does</u> fall in the critical region, we <u>reject the null hypothesis</u>. There is evidence that at least one of the dorms has a significantly different GPA.

7.   By rejecting the null hypothesis in the previous problem, we could just say that one of the population means is different. But by looking at the three different category means, it appears that dorm 2 would benefit from the new program.

8.

| | Wesaloosa Dorm | Usaloosa Dorm | Alsaloosa Dorm |
|---|---|---|---|
| | 4 | 4 | 4 |
| | 1 | 6 | 1 |
| | 0 | 5 | 8 |
| | 5 | 8 | 0 |
| | 2 | 0 | 0 |
| | 1 | 6 | 3 |
| | 3 | 2 | 4 |
| | 4 | 0 | 2 |
| | 1 | 5 | 1 |
| | 2 | 1 | 8 |
| | 0 | 6 | 6 |
| | 4 | 8 | 5 |
| | 3 | 4 | 4 |
| | 8 | 4 | 1 |
| $\sum X_i =$ | 38 | 69 | 57 |
| $\sum X^2 =$ | 166 | 439 | 369 |
| $\overline{X}_k =$ | 2.71 | 4.93 | 4.07 |
| $\overline{X} =$ | 3.90 | | |

$$SST = \sum X^2 - N\left(\overline{X}\right)^2 = (166 + 439 + 369) - (42)(3.90)^2 = 335.18$$

$$SSB = \sum N_k\left(\overline{X}_k - \overline{X}\right)^2 = (14)(2.71 - 3.90)^2 + (14)(4.93 - 3.90)^2 + (14)(4.07 - 3.90)^2$$
$$= 35.08$$

$$SSW = SST - SSB = 335.18 - 35.08 = 300.10$$

$$MSB = \frac{SSB}{k-1} = \frac{35.08}{3-1} = 17.54 \qquad MSW = \frac{SSW}{N-k} = \frac{300.10}{42-3} = 7.69$$

$$F = \frac{MSB}{MSW} = \frac{17.54}{7.69} = 2.28$$

Step 1: Make assumptions and meet test requirements.

We were instructed to assume that all test assumptions/requirements (independent random samples, interval-ratio level of measurement, normal population distribution, equal population variances) have been met.

Step 2: State the null (and research) hypotheses.

$H_o$: $\mu_1 = \mu_2 = \mu_3$

($H_1$: at least one of the population means is different)

Step 3: Select the sampling distribution and establish the critical region.
We always use the F distribution for analyses of variance. For α=0.05 at dfw=39 and dfb=2, F(critical) equals 3.32.

Step 4: Compute the test statistic (see calculations above).
F(obtained)=2.28

Step 5: Make a decision and interpret the results of the test.
Because 2.28 does <u>not</u> fall in the critical region, we <u>fail to reject the null hypothesis</u>. There is no evidence of a significant difference on the number of dates across the three dorms.

9.

| | Rural | Urban | Suburban |
|---|---|---|---|
| | 0 | 4 | 12 |
| | 3 | 6 | 0 |
| | 1 | 5 | 1 |
| | 0 | 8 | 0 |
| | 1 | 1 | 3 |
| | 5 | 0 | 4 |
| | 12 | 4 | 1 |
| | | 2 | 6 |

| | Rural | Urban | Suburban |
|---|---|---|---|
| $\sum X_i =$ | 22 | 34 | 35 |
| $\sum X^2 =$ | 180 | 172 | 257 |
| $\overline{X}_k =$ | 3.14 | 3.4 | 3.5 |
| $\overline{X} =$ | 3.37 | | |

$$SST = \sum X^2 - N(\overline{X})^2 = (180 + 172 + 257) - (27)(3.37)^2 = 302.28$$

$$SSB = \sum N_k (\overline{X}_k - \overline{X})^2 = (7)(3.14 - 3.37)^2 + (10)(3.4 - 3.37)^2 + (10)(3.5 - 3.37)^2$$
$$= 0.548$$

$$SSW = SST - SSB = 302.28 - 0.548 = 301.812$$

$$MSB = \frac{SSB}{k-1} = \frac{0.548}{3-1} = 0.274 \qquad MSW = \frac{SSW}{N-k} = \frac{302.28}{27-3} = 12.6$$

$$F = \frac{MSB}{MSW} = \frac{0.274}{12.6} = 0.022$$

**Step 1: Make assumptions and meet test requirements.**
We were instructed to assume that all test assumptions/requirements (independent random samples, interval-ratio level of measurement, normal population distribution, equal population variances) have been met.

**Step 2: State the null (and research) hypotheses.**
$H_o$: $\mu_1 = \mu_2 = \mu_3$
($H_1$: at least one of the population means is different)

**Step 3: Select the sampling distribution and establish the critical region.**
We always use the F distribution for analyses of variance. For $\alpha = 0.05$ at dfw=24 and dfb=2, F(critical) equals 3.40.

**Step 4: Compute the test statistic (see calculations above).**
F(obtained)=0.022

**Step 5: Make a decision and interpret the results of the test.**
Because 0.022 does <u>not</u> fall in the critical region, we <u>fail to reject the null hypothesis</u>. There is no evidence of a significant difference in the number of social events across the three residence areas.

10.

| | Weight Losers | Slim Quik | Forget Fat | No Dishin' | Pound No More |
|---|---|---|---|---|---|
| | 5 | 5 | 20 | 2 | 10 |
| | 30 | 10 | 21 | 5 | 14 |
| | 29 | 8 | 16 | 4 | 12 |
| | 26 | 6 | 15 | 7 | 8 |
| | 25 | 2 | 12 | 10 | 4 |
| | 10 | 10 | 8 | 6 | 7 |
| | 8 | 8 | 4 | 7 | 5 |
| | 7 | 7 | 31 | 8 | 22 |
| | 5 | 4 | 15 | 3 | 16 |
| | 22 | 12 | 10 | 2 | 5 |
| $\sum X_i =$ | 162 | 72 | 152 | 54 | 103 |
| $\sum X^2 =$ | 3789 | 609 | 2832 | 346 | 1359 |
| $\overline{X}_k =$ | 16.7 | 7.2 | 15.2 | 5.4 | 10.3 |
| $\overline{X} =$ | 10.96 | | | | |

$$SST = \sum X^2 - N\left(\overline{X}\right)^2 = (3789 + 602 + 2832 + 346 + 1359) - (50)(10.96)^2 = 2921.9$$

$$SSB = (10)(16.7 - 10.96)^2 + (10)(7.2 - 10.96)^2 + (10)(15.2 - 10.96)^2$$
$$+ (10)(5.4 - 10.69)^2 + (10)(10.3 - 10.69)^2 = 964.16$$

$$SSW = SST - SSB = 2921.9 - 964.16 = 1957.74$$

$$MSB = \frac{SSB}{k-1} = \frac{964.16}{5-1} = 241.03 \qquad MSW = \frac{SSW}{N-k} = \frac{1957.74}{50-5} = 43.51$$

$$F = \frac{MSB}{MSW} = \frac{241.03}{43.51} = 5.54$$

Step 1: Make assumptions and meet test requirements.
  We were instructed to assume that all test assumptions/requirements (independent random samples, interval-ratio level of measurement, normal population distribution, equal population variances) have been met.

Step 2: State the null (and research) hypotheses.
  $H_o$: $\mu_1 = \mu_2 = \mu_3 = \mu_4 = \mu_5$
  ($H_1$: at least one of the population means is different)

Step 3: Select the sampling distribution and establish the critical region.
  We always use the F distribution for analyses of variance. For $\alpha = 0.05$ at dfw=45 and dfb=4, F(critical) equals 2.61.

Step 4: Compute the test statistic (see calculations above).
  F(obtained)=5.54

Step 5: Make a decision and interpret the results of the test.
  Because 5.54 does fall in the critical region, we reject the null hypothesis. There is evidence of a significant difference in weight lost across the different weight-loss programs.

11.

| | Inner City | Suburban | Urban | Rural |
|---|---|---|---|---|
| | 4 | 0 | 7 | 0 |
| | 0 | 7 | 8 | 0 |
| | 8 | 6 | 9 | 5 |
| | 12 | 4 | 11 | 1 |
| | 24 | 12 | 1 | 3 |
| | 52 | 52 | 5 | 2 |
| | 100 | 8 | 0 | 4 |
| | 14 | 18 | 0 | 6 |
| | 87 | 24 | 4 | 0 |
| | 30 | 1 | 6 | 1 |
| | 2 | 6 | 3 | 3 |
| | 1 | 5 | 8 | 4 |
| | 18 | 10 | 1 | 6 |
| $\sum X_i =$ | 352 | 153 | 63 | 39 |
| $\sum X^2 =$ | 22498 | 4075 | 467 | 153 |
| $\overline{X}_k =$ | 27.1 | 11.77 | 4.85 | 3.0 |
| $\overline{X} =$ | 11.67 | | | |

$$SST = \sum X^2 - N\left(\overline{X}\right)^2 = (22498 + 4075 + 467 + 153) - (52)(11.67)^2 = 20111.1$$

$$SSB = \sum N_k\left(\overline{X}_k - \overline{X}\right)^2$$
$$= (13)(27.1 - 11.67)^2 + (13)(11.77 - 11.67)^2 + (13)(4.85 - 11.67)^2 + (13)(3.0 - 11.67)^2$$
$$= 4677.27$$

$$SSW = SST - SSB = 20111.1 - 4677.27 = 15433.83$$

$$MSB = \frac{SSB}{k-1} = \frac{4677.27}{4-1} = 1559.09 \qquad MSW = \frac{SSW}{N-k} = \frac{15433.83}{52-4} = 321.54$$

$$F = \frac{MSB}{MSW} = \frac{1559.09}{321.54} = 4.85$$

Step 1: Make assumptions and meet test requirements.

We were instructed to assume that all test assumptions/requirements (independent random samples, interval-ratio level of measurement, normal population distribution, equal population variances) have been met.

Step 2: State the null (and research) hypotheses.

H$_0$: $\mu_1 = \mu_2 = \mu_3 = \mu_4$

(H$_1$: at least one of the population means is different)

Step 3: Select the sampling distribution and establish the critical region.
We always use the F distribution for analyses of variance. For α=0.05 at dfw=48 and dfb=3, F(critical) equals 2.84.

Step 4: Compute the test statistic (see calculations above).
F(obtained)=4.85

Step 5: Make a decision and interpret the results of the test.
Because 4.85 <u>does</u> fall in the critical region, we <u>reject the null hypothesis</u>.
There is evidence of a significant difference in the calls to the police across these places of residence.

12.    When alpha is changed to 0.01, the only parts of the ANOVA test to change are steps 3 and 5.

Step 3: Select the sampling distribution and establish the critical region.
We always use the F distribution for analyses of variance. For α=0.05 at dfw=48 and dfb=3, F(critical) equals 4.31.

Step 5: Make a decision and interpret the results of the test.
Because 4.85 <u>does</u> fall in the critical region, we <u>reject the null hypothesis</u>.
There is evidence of a significant difference in the calls to the police across these places of residence.

Yes, there is still a significant difference across the populations.

13.

| dfw | dfb | F(critical) α=0.05 | F(critical) α=0.01 |
|-----|-----|--------------------|--------------------|
| 1   | 1   | 161.4              | 4052               |
| 10  | 4   | 3.48               | 5.99               |
| 35  | 10  | 2.27               | 3.17               |
| 100 | 5   | 2.37               | 3.34               |
| 58  | 3   | 2.85               | 4.31               |

## SPSS or MicroCase WORK PROBLEMS

If you had problems with any of these questions, be sure to re-read the appropriate sections at the end of the chapter in your textbook.

1.  The null hypothesis states that there is no difference among these populations for the number of children. To test this, we need to conduct an ANOVA test comparing the mean number of children across the five populations. From the output below we see that those with less than a high school diploma have the most children (mean=2.74), and those with a graduate degree have the fewest children (mean=1.29. To determine if any of these differences are significant, the ANOVA test is used. The F statistic of 21.088 has a significance level of 0.000. Therefore, we reject the null hypothesis; the evidence suggests that at least one of these groups is significantly different from the others.

**Descriptives**

| | N | Mean | Std. Deviation | Std. Error | 95% Confidence Interval for Mean | | Minimum | Maximum |
|---|---|---|---|---|---|---|---|---|
| | | | | | Lower Bound | Upper Bound | | |
| LT HIGH SCHOOL | 221 | 2.74 | 2.056 | .138 | 2.47 | 3.01 | 0 | 8 |
| HIGH SCHOOL | 767 | 1.89 | 1.568 | .057 | 1.77 | 2.00 | 0 | 8 |
| JUNIOR COLLEGE | 114 | 1.87 | 1.588 | .149 | 1.57 | 2.16 | 0 | 7 |
| BACHELOR | 262 | 1.66 | 1.538 | .095 | 1.47 | 1.84 | 0 | 6 |
| GRADUATE | 133 | 1.29 | 1.277 | .111 | 1.07 | 1.50 | 0 | 6 |
| Total | 1497 | 1.92 | 1.666 | .043 | 1.83 | 2.00 | 0 | 8 |

**ANOVA**

| | Sum of Squares | df | Mean Square | F | Sig. |
|---|---|---|---|---|---|
| Between Groups | 222.273 | 4 | 55.568 | 21.088 | .000 |
| Within Groups | 3931.455 | 1492 | 2.635 | | |
| Total | 4153.729 | 1496 | | | |

2.  Those with less than a high school diploma watch the most television, averaging 3.78 hours per day, and those with a graduate degree watch the least amount of television per day (mean=2.00). The F statistic (7.337) is significant at the 0.000 level. We reject the null hypothesis of equal population means and conclude that at least one of the populations is significantly different. (See next page.)

**Descriptives**

| | N | Mean | Std. Deviation | Std. Error | 95% Confidence Interval for Mean | | Minimum | Maximum |
|---|---|---|---|---|---|---|---|---|
| | | | | | Lower Bound | Upper Bound | | |
| LT HIGH SCHOOL | 88 | 3.78 | 2.739 | .292 | 3.20 | 4.36 | 0 | 15 |
| HIGH SCHOOL | 340 | 3.14 | 2.128 | .115 | 2.91 | 3.37 | 0 | 14 |
| JUNIOR COLLEGE | 58 | 2.57 | 1.902 | .250 | 2.07 | 3.07 | 0 | 9 |
| BACHELOR | 129 | 2.54 | 2.828 | .249 | 2.05 | 3.04 | 0 | 24 |
| GRADUATE | 58 | 2.00 | 1.487 | .195 | 1.61 | 2.39 | 0 | 6 |
| Total | 673 | 2.96 | 2.346 | .090 | 2.79 | 3.14 | 0 | 24 |

**ANOVA**

| | Sum of Squares | df | Mean Square | F | Sig. |
|---|---|---|---|---|---|
| Between Groups | 155.710 | 4 | 38.928 | 7.337 | .000 |
| Within Groups | 3544.361 | 668 | 5.306 | | |
| Total | 3700.071 | 672 | | | |

3. Those with less than a high school diploma have the lowest income ($15,647.10) and those with a graduate degree have the highest income ($58,450.20). The F statistic (61.751) is significant at the 0.000 level. We reject the null hypothesis of equal population means and conclude that at least one of the populations is significantly different.

**Descriptives**

| | N | Mean | Std. Deviation | Std. Error | 95% Confidence Interval for Mean | | Minimum | Maximum |
|---|---|---|---|---|---|---|---|---|
| | | | | | Lower Bound | Upper Bound | | |
| LT HIGH SCHOOL | 180 | 15647.1 | 17849.5 | 1330.42 | 13021.8 | 18272.5 | 275 | 139981 |
| HIGH SCHOOL | 654 | 27067.5 | 23141.9 | 904.92 | 25290.6 | 28844.4 | 275 | 139981 |
| JUNIOR COLLEGE | 98 | 40024.4 | 34372.1 | 3472.11 | 33133.2 | 46915.6 | 3579 | 139981 |
| BACHELOR | 233 | 47597.9 | 37378.5 | 2448.75 | 42773.2 | 52422.5 | 1101 | 139981 |
| GRADUATE | 117 | 58450.2 | 44256.6 | 4091.52 | 50346.4 | 66554.0 | 1101 | 139981 |
| Total | 1282 | 33049.9 | 31612.5 | 882.91 | 31317.8 | 34782.0 | 275 | 139981 |

**ANOVA**

| | Sum of Squares | df | Mean Square | F | Sig. |
|---|---|---|---|---|---|
| Between Groups | 2.07E+11 | 4 | 5.187E+10 | 61.751 | .000 |
| Within Groups | 1.07E+12 | 1277 | 840003311.1 | | |
| Total | 1.28E+12 | | | | |

231

4.  Those with less than a high school diploma have the lowest prestige and those with a graduate degree have the highest prestige. The F statistic (234.058) is significant at the 0.000 level. We reject the null hypothesis of equal population means and conclude that at least one of the populations is significantly different.

**Descriptives**

| | N | Mean | Std. Deviation | Std. Error | 95% Confidence Interval for Mean | | Minimum | Maximum |
|---|---|---|---|---|---|---|---|---|
| | | | | | Lower Bound | Upper Bound | | |
| LT HIGH SCHOOL | 199 | 34.023 | 11.816 | .838 | 32.37 | 35.68 | 17.1 | 87.0 |
| HIGH SCHOOL | 725 | 43.840 | 15.477 | .575 | 42.71 | 44.97 | 17.1 | 93.3 |
| JUNIOR COLLEGE | 110 | 56.645 | 16.951 | 1.616 | 53.44 | 49.85 | 27.5 | 86.9 |
| BACHELOR | 261 | 63.492 | 16.765 | 1.038 | 61.45 | 65.54 | 22.5 | 93.3 |
| GRADUATE | 131 | 76.090 | 14.734 | 1.287 | 73.54 | 78.64 | 22.5 | 97.2 |
| Total | 1426 | 50.017 | 19.709 | .522 | 48.99 | 51.04 | 17.1 | 97.2 |

**ANOVA**

| | Sum of Squares | df | Mean Square | F | Sig. |
|---|---|---|---|---|---|
| Between Groups | 219850.0 | 4 | 54962.51 | 234.058 | .000 |
| Within Groups | 333685.4 | 1421 | 234.824 | | |
| Total | 553535.4 | 1425 | | | |

# CHAPTER ELEVEN
## Hypothesis Testing IV: Chi Square

_Learning Objectives_: _By the end of this chapter, you will be able to:_

1. _Identify and cite examples of situations in which the chi square test is appropriate._
2. _Explain the structure of a bivariate table and the concept of independence as applied to expected and observed frequencies in a bivariate table._
3. _Explain the logic of hypothesis testing as applied to a bivariate table._
4. _Perform the chi square test using the five-step model and correctly interpret the results._
5. _Explain the limitations of the chi square test and, especially, the difference between statistical significance and importance._

## KEY TERMS

- ✓ Bivariate Table  (p. 256)
- ✓ Cells  (p. 257)
- ✓ $\chi^2$ (Critical)  (p. 259)
- ✓ $\chi^2$ (Obtained)  (p. 259)
- ✓ Chi Square Test  (p. 256)
- ✓ Column  (p. 257)
- ✓ Expected Frequency ($f_e$)  (p. 259)
- ✓ Independence  (p. 258)
- ✓ Marginals  (p. 257)
- ✓ Nonparametric  (p. 256)
- ✓ Observed Frequency ($f_o$)  (p. 259)
- ✓ Row  (p. 257)

## CHAPTER SUMMARY

The past few chapters have introduced you to various types of hypotheses tests. Chapter eleven introduces the Chi Square test, one of the most frequently used hypothesis tests in the social sciences. The Chi Square test is popular because it's assumptions are easily met and because it can be used with variables measured at any level, including the nominal level.

# CHAPTER OUTLINE

## 11.1 Introduction.

- One of the most commonly used statistical tests in the social sciences is the chi square test ($\chi^2$) of independence. Chi square is a non-parametric statistic, which means that no assumptions are made about the shape of the distribution. This makes chi square very popular. Another reason for its popularity is its ability to be used with nominal data.

## 11.2 Bivariate Tables.

- Chi square is especially easy to use when you have small sample sizes and your data are arranged in bivariate tables. We usually put the independent variable (the "cause") in the columns and the dependent variable (the "effect") in the rows of the bivariate table. Each cell of the table includes the combined frequency on both variables.

## 11.3 The Logic of Chi Square.

- In the chi square test, the null hypothesis is that the two variables are independent of one another, meaning that being placed in one of categories of the first variable is independent of being placed in any of the categories of the second variable. Think of it like this: if we have two groups, males and females, and the variable attitude toward abortion rights (pro-choice or pro-life), we can say that the two variables (gender and attitude toward abortion rights) are independent if the chance of being in either category of the dependent variable (pro-choice or pro-life) is not affected by gender. To put it another way: being male or female would not affect the chance of being either pro-choice or pro-life. If the two variables are dependent on one another, then being male or female would affect being either pro-choice or pro-life.

## 11.4 The Computation of Chi Square.

- The null hypothesis is always that the two variables (for example, gender and attitude toward abortion rights) are independent of one another. The research hypothesis, then, is that they are dependent.
- The chi square test uses the row and column marginals (totals) to create expected frequencies ($f_e$), or the cell frequencies we would expect to get if the two variables were independent of each other. These expected frequencies are then compared to the observed frequencies ($f_o$), or the cell frequencies that are actually observed with our data. If there is a large difference between the expected and observed frequencies, then we conclude that the two variables are dependent on one another and reject the null hypothesis.

## 11.5 The Chi Square Test for Independence.

- Let's review the five step model for the chi square hypothesis test. State the assumptions and test requirements (independent random samples and nominal level of measurement) in <u>step 1</u>, and state the hypotheses in <u>step 2</u>. Select the sampling distribution (chi square) and establish the critical region using the degrees of freedom $[df=(r-1)(c-1)]$ and Appendix C in <u>step 3</u>. Compute the test statistic, $\chi^2$(obtained) in <u>step 4</u>. Make a decision and interpret the results in <u>step 5</u>.

## 11.6 The Chi Square Test: An Additional Example.

## 11.7 The Limitations of the Chi Square Test.

- Like other hypothesis tests, the chi square test can only tell us that the variables are probably related in the population (for example, gender and abortion attitudes). It cannot tell us *how* the two variables are related to each other. (For example, it cannot tell us if women are more likely to be pro-choice than men are.) In order to determine *how* the variables are related, we calculate column percents [$100 \times$ (cell frequency/column total)] and then compare those column percents within categories of the dependent (row) variable (that is, we compare the column percents across the row).
- The chi square statistic is affected by sample size; in fact, the chi square statistic increases at the same rate as the sample size. Thus, if you have a large sample, very small differences will be exaggerated, and trivial differences may turn out to be significant. In addition, you should use caution whenever you have any cells with frequencies less than five. The chi square statistic cannot be adequately computed with such small cell sizes, although you can correct for this problem on 2x2 tables with Yate's correction for continuity (see formula 10.4 in your textbook). Last, chi square becomes difficult to interpret when the variables have many categories. Your textbook recommends using chi square when both variables have four or fewer categories.

## MULTIPLE-CHOICE QUESTIONS

1.  In the null hypothesis statement for the chi square test, the two variables are hypothesized to be _____.
    a. dependent
    b. independent
    c. associated
    d. equal variances

*Learning Objective: 2*

2. Row marginals are _____.
   a. the total sample size
   b. the vertical subtotals
   c. the horizontal subtotals
   d. the largest cell frequency

   *Learning Objective: 2*

3. Observed frequencies are _____.
   a. cell frequencies expected if the two variables are independent
   b. total frequencies observed
   c. total frequencies expected
   d. cell frequencies observed from the raw data

   *Learning Objective: 2*

4. Expected frequencies are _____.
   a. cell frequencies expected if the two variables are independent
   b. total frequencies observed
   c. total frequencies expected
   d. cell frequencies observed from the raw data

   *Learning Objective: 2*

5. _____ frequencies are the raw data, whereas _____ frequencies are found with row and column marginals.
   a. Expected, observed
   b. Total, expected
   c. Observed, total
   d. Observed, expected

   *Learning Objective: 2*

6. In order to find the observed frequencies _____.
   a. multiply the row and column marginals and divide by the sample size
   b. add up all of the frequencies for each cell from the raw data
   c. divide the cell frequencies by the total sample size
   d. multiply the sample size by each cell frequency

   *Learning Objective: 2*

7.    In order to find the expected frequencies _____.
      a.  multiply the row and column marginals and divide by the sample size
      b.  add up all of the frequencies for each cell from the raw data
      c.  divide the cell frequencies by the total sample size
      d.  multiply the sample size by each cell frequency

*Learning Objective: 2*


8.    In order to test for independence, _____ is/are compared to _____.
      a.  chi square obtained, chi square critical
      b.  row marginals, column marginals
      c.  degrees of freedom, cell frequencies
      d.  marginal totals, sample size

*Learning Objective: 4*


9.    Degrees of freedom are found with which of the following formulas?
      a.  (number of rows) (sample size)
      b.  (number of rows)(number of columns)
      c.  (number of rows-1)(number of columns-1)
      d.  (sample size, N) (number of cells)

*Learning Objective: 4*


10.   In order to find the critical value in the chi square test, you need _____ and _____.
      a.  row marginals, sample size
      b.  degrees of freedom, critical value
      c.  sample size, column marginals
      d.  degrees of freedom, alpha level

*Learning Objective: 4*


11.   The chi square test is affected by _____.
      a.  row marginals
      b.  sample size
      c.  cell frequencies
      d.  all of these choices

*Learning Objective: 5*

12. The chi square critical value for a five by four matrix, and an alpha level of 0.01 is
_____.
   a. 3.571
   b. 12
   c. 21.026
   d. 26.217

*Learning Objective: 4*


13. If a chi square test statistic was calculated to be 4.72, with two degrees of freedom, it would be rejected at _____ alpha level.
   a. 0.001
   b. 0.01
   c. 0.05
   d. 0.10

*Learning Objective: 4*


14. In order to reject a null hypothesis, the obtained value must fall _____ the critical value.
   a. within
   b. to the left of
   c. to the right of
   d. nowhere near

*Learning Objective: 3*


15. The sampling distribution of the chi square is _____ .
   a. positively skewed
   b. negatively skewed
   c. normal in shape
   d. determined by the means of the population

*Learning Objective: 3*

16. Which of the below is not one of the reasons why the chi square test has probably been the most frequently used test of hypothesis?
   a. the assumptions are easy to satisfy
   b. it can be conducted on nominal variables
   c. we don't need to assume anything about the shape of the sampling distribution
   d. it can only be used with interval-ratio variables

   *Learning Objective: 1*


17. By convention, in a bivariate table the _____ variable is put in the rows, and the _____ variable is put in the columns.
   a. independent, dependent
   b. dependent, independent
   c. nominal, ordinal
   d. ordinal, nominal

   *Learning Objective: 2*


18. Two variables are said to be _____ if the classification of a case into a particular category of the independent variable has no effect on the probability that the case will fall into any particular category of the dependent variable.
   a. dependent
   b. related
   c. independent
   d. significant

   *Learning Objective: 3*


19. If the null hypothesis of a chi square hypothesis test has been rejected, then _____.
   a. there should be large differences between observed and expected frequencies
   b. there should be small to no differences between observed and expected frequencies
   c. the variables are statistically independent
   d. the variables are not related in the population

   *Learning Objective: 3*

20. Which of the below is the formula for calculating $\chi^2$(obtained).

a. $\sum \dfrac{(f_e - f_o)^2}{f_e}$

b. $\sum \dfrac{(f_o - f_e)}{f_e}$

c. $\sum \dfrac{(f_o - f_e)^2}{f_e}$

d. $\sum \dfrac{(f_e - f_o)}{f_o}$

*Learning Objective: 4*

## WORK PROBLEMS

For all of the work problems, assume that the test assumptions and requirements have been met. Use an alpha level of 0.05, unless stated otherwise.

1. The following data show the distribution of job satisfaction and life satisfaction. Using only the one expected frequency, can you find the other three with subtraction?

| ($F_o$) | Job Satisfaction | | |
|---|---|---|---|
| Life Satisfaction | High | Low | Total |
| High | 54 | 29 | 83 |
| Low | 31 | 65 | 96 |
| Total | 85 | 94 | 179 |

| ($F_e$) | Job Satisfaction | | |
|---|---|---|---|
| Life Satisfaction | High | Low | Total |
| High | 39.4 | | 83 |
| Low | | | 96 |
| Total | 85 | 94 | 179 |

*Learning Objective: 2*

2. From the data above, what can we say about the two variables, job satisfaction and life satisfaction? Are the two variables independent of one another?

*Learning Objective: 4*

3.  For the data below, find the expected frequencies using the subtraction method and only those cell frequencies given. Note: because this is a two by three matrix, you need only two pieces of information to complete the matrix, therefore the degrees of freedom is two.

| ($F_e$) | Dormitory | | | |
|---|---|---|---|---|
| | Dorm 1 | Dorm 2 | Dorm 3 | Total |
| Favor | 17.9 | | | 51 |
| Oppose | | 14.7 | | 49 |
| Total | 35 | 30 | 35 | 100 |

*Learning Objective: 2*

4.  The following data were collected to determine if gender and marital happiness were related. Calculate the expected frequencies using the formula from your text.

| ($F_o$) | Gender | | |
|---|---|---|---|
| Marital Satisfaction | Male | Female | Total |
| Very Happy | 30 | 45 | 75 |
| Moderately Happy | 50 | 45 | 95 |
| Unhappy | 20 | 10 | 30 |
| Total | 100 | 100 | 200 |

*Learning Objective: 2*

5.  Based on the previous problem how many expected frequencies would you have needed in order to use the subtraction method for determining the rest?

*Learning Objective: 2*

6.  Using the data from problem 4, determine if the two variables are independent of one another?

*Learning Objective: 4*

7. A researcher suspected that attitudes toward gun control were dependent on gender. Do the data below support his hypothesis?

| (F$_o$) | Gender | | |
|---|---|---|---|
| Attitude Toward Gun Control | Male | Female | Total |
| Favor | 45 | 42 | 87 |
| Oppose | 39 | 49 | 88 |
| Total | 84 | 91 | 175 |

*Learning Objective: 4*


8. Knowledge of birth control was measured for men and women. Are the two variables independent of one another? What problems do we encounter in this test?

| (F$_o$) | Gender | | |
|---|---|---|---|
| Knowledge of Birth Control | Male | Female | Total |
| High | 170 | 430 | 600 |
| Moderate | 510 | 220 | 730 |
| Low | 210 | 160 | 370 |
| Total | 890 | 810 | 1700 |

*Learning Objective: 4 & 5*


9. Divide each of the observed frequencies and marginals in the previous problem by ten. Are the two variables independent?

*Learning Objective: 4 & 5*


10. With more women in the work force these days, a sociologist wondered whether men were taking over more household responsibilities. She collected the following data. Are gender and housework independent?

| (F$_o$) | Gender | | |
|---|---|---|---|
| Amount of Housework | Male | Female | Total |
| All | 15 | 46 | 61 |
| Most | 20 | 53 | 73 |
| Some | 69 | 27 | 96 |
| None | 11 | 9 | 20 |
| Total | 115 | 135 | 250 |

*Learning Objective: 4*

11. Crime is perceived to be a greater threat in the city than in the suburbs. A criminologist collected data on the victimization of crimes for both areas. Do the data below indicate that the two variables are dependent?

| $(F_o)$ Victim of Crime | Area of Residence | | |
|---|---|---|---|
| | Urban | Suburban | Total |
| Yes | 40 | 25 | 65 |
| No | 25 | 35 | 60 |
| Total | 65 | 60 | 125 |

*Learning Objective: 4*

12. What is the critical value for a chi square test with five degrees of freedom at the 0.01 level?

*Learning Objective: 4*

13. If a researcher found a chi square obtained of 6.75, in a test with three degrees of freedom, would she be able to reject the null hypothesis at the 0.01 level? If not, at what level could she reject the null hypothesis?

*Learning Objective: 4*

14. In a poll for the approval rating of a new Democratic president, a newspaper showed the following data. Is the approval rating dependent on political party preference?

| $(F_o)$ Approval of President | Political Party Preference | | |
|---|---|---|---|
| | Republican | Democrat | Total |
| Yes | 14 | 29 | 43 |
| No | 24 | 11 | 35 |
| Total | 38 | 40 | 78 |

*Learning Objective: 4*

15. In a similar poll, respondents were asked to rate their politics as either liberal or conservative. Are these labels independent of political party?

| (F$_o$) Political Party Affiliation | Gender | | Total |
|---|---|---|---|
| | Liberal | Conservative | |
| Republican | 3 | 25 | 28 |
| Democrat | 29 | 2 | 31 |
| Independent | 10 | 8 | 18 |
| Total | 42 | 35 | 77 |

*Learning Objective: 4*

16. A newly elected congressman is considering introducing a new bill to legalize marijuana, but first wants to know if gender plays a role in the attitude toward the issue. Are the two variables independent?

| (F$_o$) Legalization of Marijuana | Gender | | Total |
|---|---|---|---|
| | Male | Female | |
| Approve | 10 | 6 | 16 |
| Disapprove | 5 | 9 | 14 |
| No Opinion | 9 | 11 | 20 |
| Total | 24 | 26 | 50 |

*Learning Objective: 4*

17. Bob Jherque, the new mayor, has proposed the construction of a new water plant. Is the approval rating dependent on whether you live in town or on a farm?

| (F$_o$) Approval of Water Plant | Place of Residence | | Total |
|---|---|---|---|
| | Town | Farm | |
| Yes | 33 | 46 | 79 |
| No | 37 | 30 | 67 |
| Total | 70 | 76 | 146 |

*Learning Objective: 4*

**SPSS or MicroCase WORK PROBLEMS**

The following work problems can be completed using either SPSS or MicroCase. The solutions will be the same regardless of which one you use. Your instructor may have a preference, so be sure to check with him/her prior to beginning these problems.

1.      Are the variables gender (*sex*) and attitudes toward abortion (*abany*) independent of one another? Using the GSS 2006 dataset, test this research question.

2.      Are the variables gender (*sex*) and attitudes toward extramarital sex (*xmarsex*) independent of one another?

3.      Do married (*marital*) people report being happier (*happy*) than non-married people? Are these two variables independent or dependent on one another?

4.      Are the variables gender (*sex*) and attitudes toward sex education (*sexeduc*) independent of one another?

5.      Are the variables gender (*sex*) and attitudes toward premarital sex (*premarsx*) independent of one another?

**WORK PROBLEMS**

1.

| (F_e) | Job Satisfaction | | |
|---|---|---|---|
| Life Satisfaction | High | Low | Total |
| High | 39.4 | 43.6 | 83 |
| Low | 45.6 | 50.4 | 96 |
| Total | 85 | 94 | 179 |

2.

$$\chi^2(obtained) = \sum \frac{(f_o - f_e)^2}{f_e} =$$

$$= \frac{(54 - 39.4)^2}{39.4} + \frac{(29 - 43.6)^2}{43.6} + \frac{(31 - 45.6)^2}{45.6} + \frac{(65 - 50.4)^2}{50.4} = 19.2$$

Step 1: Make assumptions and meet test requirements.
We were instructed to assume that all test assumptions/requirements (independent random samples and nominal level of measurement) have been met.

Step 2: State the null (and research) hypotheses.
$H_o$: The two variables are independent ($\chi^2=0$)
($H_1$: The two variables are dependent)

Step 3: Select the sampling distribution and establish the critical region.
For $\alpha=0.05$ at df=(r-1)(c-1)=(2-1)(2-1)=1, $\chi^2$(critical) equals 3.841.

Step 4: Compute the test statistic (see calculations above).
$\chi^2$(obtained)=19.2

Step 5: Make a decision and interpret the results of the test.
Because 19.2 <u>does</u> fall in the critical region, we <u>reject the null hypothesis</u> of independence and conclude that job satisfaction and life satisfaction are dependent.

3.

| (F_e) | Dormitory | | | |
|---|---|---|---|---|
| | Dorm 1 | Dorm 2 | Dorm 3 | Total |
| Favor | 17.9 | 15.3 | 17.8 | 51 |
| Oppose | 17.1 | 14.7 | 17.2 | 49 |
| Total | 35 | 30 | 35 | 100 |

4.

| (F_e) | Gender | | |
| Marital Satisfaction | Male | Female | Total |
| --- | --- | --- | --- |
| Very Happy | 37.5 | 37.5 | 75 |
| Moderately Happy | 47.5 | 47.5 | 95 |
| Unhappy | 15 | 15 | 30 |
| Total | 100 | 100 | 200 |

5.    $df = (r - 1)(c - 1) = (2 - 1)(3 - 1) = 2$

A table with two columns and three rows has 2 degrees of freedom, so we would have to have been given two expected frequencies to use the subtraction method to determine the rest.

6.

$$\chi^2(obtained) = \frac{(30 - 37.5)^2}{37.5} + \frac{(45 - 37.5)^2}{37.5} + \frac{(50 - 47.5)^2}{47.5} + \frac{(45 - 47.5)^2}{47.5}$$

$$+ \frac{(20 - 15)^2}{15} + \frac{(10 - 15)^2}{15} = 6.66$$

Step 1: Make assumptions and meet test requirements.
We were instructed to assume that all test assumptions/requirements (independent random samples and nominal level of measurement) have been met.

Step 2: State the null (and research) hypotheses.
H_o: The two variables are independent ($\chi^2=0$)
(H_1: The two variables are dependent)

Step 3: Select the sampling distribution and establish the critical region.
For $\alpha=0.05$ at $df=(r-1)(c-1)=(3-1)(2-1)=2$, $\chi^2$(critical) equals 5.991.

Step 4: Compute the test statistic (see calculations above).
$\chi^2$(obtained)=6.66

Step 5: Make a decision and interpret the results of the test.
Because 6.66 <u>does</u> fall in the critical region, we <u>reject the null hypothesis</u> of independence and conclude that marital satisfaction and gender are dependent.

7.

| (Fe) | Gender | | |
|---|---|---|---|
| Attitude Toward Gun Control | Male | Female | Total |
| Favor | 41.8 | 45.2 | 87 |
| Oppose | 42.2 | 45.8 | 88 |
| Total | 84 | 91 | 175 |

$$\chi^2 (obtained) = \frac{(41-41.8)^2}{41.8} + \frac{(42-45.2)^2}{45.2} + \frac{(39-42.2)^2}{42.2} + \frac{(49-45.8)^2}{45.8} = 0.96$$

Step 1: Make assumptions and meet test requirements.
We were instructed to assume that all test assumptions/requirements (independent random samples and nominal level of measurement) have been met.

Step 2: State the null (and research) hypotheses.
$H_o$: The two variables are independent ($\chi^2=0$)
($H_1$: The two variables are dependent)

Step 3: Select the sampling distribution and establish the critical region.
For $\alpha=0.05$ at df=(r-1)(c-1)=(2-1)(2-1)=1, $\chi^2$(critical) equals 3.841.

Step 4: Compute the test statistic (see calculations above).
$\chi^2$(obtained)=0.96

Step 5: Make a decision and interpret the results of the test.
Because 0.96 does <u>not</u> fall in the critical region, we <u>fail to reject the null hypothesis</u> of independence and conclude that attitudes toward gun control and gender are independent.

8.

| (Fe) | Gender | | |
|---|---|---|---|
| Knowledge of Birth Control | Male | Female | Total |
| High | 314.1 | 285.9 | 600 |
| Moderate | 382.2 | 347.8 | 730 |
| Low | 193.7 | 176.3 | 370 |
| Total | 890 | 810 | 1700 |

$$\chi^2 (obtained) = \frac{(170-314.1)^2}{314.1} + \frac{(510-382.2)^2}{382.2} + \frac{(210-193.7)^2}{193.7} +$$

$$\frac{(430-285.9)^2}{285.9} + \frac{(220-347.8)^2}{347.8} + \frac{(160-176.3)^2}{176.3} = 231.3$$

Step 1: Make assumptions and meet test requirements.

We were instructed to assume that all test assumptions/requirements (independent random samples and nominal level of measurement) have been met.

Step 2: State the null (and research) hypotheses.

$H_o$: The two variables are independent ($\chi^2=0$)

($H_1$: The two variables are dependent)

Step 3: Select the sampling distribution and establish the critical region.

For $\alpha=0.05$ at df=(r-1)(c-1)=(3-1)(2-1)=2, $\chi^2$(critical) equals 5.991.

Step 4: Compute the test statistic (see calculations above).

$\chi^2$(obtained)=231.3

Step 5: Make a decision and interpret the results of the test.

Because 231.3 <u>does</u> fall in the critical region, we <u>reject the null hypothesis</u> of independence and conclude that knowledge of birth control and gender are dependent.

One problem, though, is the large sample sizes. Because the sample sizes are 890 and 810, even the smallest differences will appear significant. Therefore, although we reject the null hypothesis of independence, we do so with caution.

9.

($F_o$)

| Knowledge of Birth Control | Gender Male | Female | Total |
|---|---|---|---|
| High | 17 | 43 | 60 |
| Moderate | 51 | 22 | 73 |
| Low | 21 | 16 | 37 |
| Total | 89 | 81 | 170 |

($F_e$)

| Knowledge of Birth Control | Gender Male | Female | Total |
|---|---|---|---|
| High | 31.4 | 28.6 | 60 |
| Moderate | 38.2 | 34.8 | 73 |
| Low | 19.4 | 17.6 | 37 |
| Total | 89 | 81 | 170 |

$$\chi^2 (obtained) = \frac{(17-31.4)^2}{31.4} + \frac{(51-38.2)^2}{38.2} + \frac{(21-19.4)^2}{19.4} +$$
$$\frac{(43-28.6)^2}{28.6} + \frac{(22-34.8)^2}{34.8} + \frac{(16-17.6)^2}{17.6} = 23.1$$

Step 1: Make assumptions and meet test requirements.
　　　　We were instructed to assume that all test assumptions/requirements (independent random samples and nominal level of measurement) have been met.

Step 2: State the null (and research) hypotheses.
　　　　$H_o$:　The two variables are independent ($\chi^2=0$)
　　　　($H_1$:　The two variables are dependent)

Step 3: Select the sampling distribution and establish the critical region.
　　　　For $\alpha=0.05$ at df=(r-1)(c-1)=(3-1)(2-1)=2, $\chi^2$(critical) equals 5.991.

Step 4: Compute the test statistic (see calculations above).
　　　　$\chi^2$(obtained)=23.1

Step 5: Make a decision and interpret the results of the test.
　　　　Because 23.1 <u>does</u> fall in the critical region, we <u>reject the null hypothesis</u> of independence and conclude that knowledge of birth control and gender are dependent.

Thus, even after dividing the sample size by ten, we find a significant relationship.

10.

| ($F_e$) | Gender | | |
| Amount of Housework | Male | Female | Total |
| --- | --- | --- | --- |
| All | 28.1 | 32.9 | 61 |
| Most | 33.6 | 39.4 | 73 |
| Some | 44.2 | 51.8 | 96 |
| None | 9.2 | 10.8 | 20 |
| Total | 115 | 135 | 250 |

$$\chi^2(obtained) = \frac{(15-28.1)^2}{28.1} + \frac{(20-33.6)^2}{33.6} + \frac{(69-44.2)^2}{44.2} + \frac{(11-9.2)^2}{9.2} +$$

$$\frac{(46-32.9)^2}{32.9} + \frac{(53-39.4)^2}{39.4} + + \frac{(27-51.8)^2}{51.8} + \frac{(9-10.8)^2}{10.8} = 47.9$$

Step 1: Make assumptions and meet test requirements.
　　　　We were instructed to assume that all test assumptions/requirements (independent random samples and nominal level of measurement) have been met.

Step 2: State the null (and research) hypotheses.
　　　　$H_o$:　The two variables are independent ($\chi^2=0$)
　　　　($H_1$:　The two variables are dependent)

Step 3: Select the sampling distribution and establish the critical region.
For α=0.05 at df=(r-1)(c-1)=(4-1)(2-1)=3, $\chi^2$(critical) equals 7.815.

Step 4: Compute the test statistic (see calculations above).
$\chi^2$(obtained)=47.9

Step 5: Make a decision and interpret the results of the test.
Because 47.9 does fall in the critical region, we reject the null hypothesis of independence and conclude that amount of housework and gender are dependent.

11.

| (F$_e$) | Area of Residence | | |
|---|---|---|---|
| Victim of Crime | Urban | Suburban | Total |
| Yes | 33.8 | 31.2 | 65 |
| No | 31.2 | 28.8 | 60 |
| Total | 65 | 60 | 125 |

$$\chi^2 (obtained) = \frac{(40-33.8)^2}{33.8} + \frac{(25-31.2)^2}{31.2} + \frac{(25-31.2)^2}{31.2} + \frac{(25-28.8)^2}{28.8} = 4.93$$

Step 1: Make assumptions and meet test requirements.
We were instructed to assume that all test assumptions/requirements (independent random samples and nominal level of measurement) have been met.

Step 2: State the null (and research) hypotheses.
H$_o$:  The two variables are independent ($\chi^2$=0)
(H$_1$:  The two variables are dependent)

Step 3: Select the sampling distribution and establish the critical region.
For α=0.05 at df=(r-1)(c-1)=(2-1)(2-1)=1, $\chi^2$(critical) equals 3.841.

Step 4: Compute the test statistic (see calculations above).
$\chi^2$(obtained)=4.93

Step 5: Make a decision and interpret the results of the test.
Because 4.93 does fall in the critical region, we reject the null hypothesis of independence and conclude that crime victimization and gender are dependent.

12. The critical value for a chi square test with five degrees of freedom at the 0.01 level is 15.086.

13.   Step 1: Make assumptions and meet test requirements.
      We were instructed to assume that all test assumptions/requirements (independent random samples and nominal level of measurement) have been met.

Step 2: State the null (and research) hypotheses.
      $H_o$:  The two variables are independent ($\chi^2=0$)
      ($H_1$:  The two variables are dependent)

Step 3: Select the sampling distribution and establish the critical region.
      For $\alpha=0.01$ at df=3, $\chi^2$(critical) equals 11.341.

Step 4: Compute the test statistic (given in problem).
      $\chi^2$(obtained)=6.75

Step 5: Make a decision and interpret the results of the test.
      Because 6.75 does <u>not</u> fall in the critical region, we <u>fail to reject the null hypothesis</u> of independence and conclude that the two variables are independent.

For $\alpha=0.05$ at df=3, $\chi^2$(critical) equals 7.815, and for $\alpha=0.10$ at df=3, $\chi^2$(critical) equals 6.251.  Only with this last critical value (6.251 for $\alpha=0.10$) is the researcher able to reject the null hypothesis and conclude that the two variables are dependent.

14.

| ($F_e$) | Political Party Preference | | |
| --- | --- | --- | --- |
| Approval of President | Republican | Democrat | Total |
| Yes | 20.9 | 22.1 | 43 |
| No | 17.1 | 17.9 | 35 |
| Total | 38 | 40 | 78 |

$$\chi^2 (obtained) = \frac{(14-20.9)^2}{20.9} + \frac{(29-22.1)^2}{22.1} + \frac{(24-17.1)^2}{17.1} + \frac{(11-17.9)^2}{17.9} = 9.87$$

Step 1: Make assumptions and meet test requirements.
      We were instructed to assume that all test assumptions/requirements (independent random samples and nominal level of measurement) have been met.

Step 2: State the null (and research) hypotheses.
      $H_o$:  The two variables are independent ($\chi^2=0$)
      ($H_1$:  The two variables are dependent)

Step 3: Select the sampling distribution and establish the critical region.
      For $\alpha=0.05$ at df=(r-1)(c-1)=(2-1)(2-1)=1, $\chi^2$(critical) equals 3.841.

Step 4: Compute the test statistic (see calculations above).
$\chi^2$(obtained)=9.87

Step 5: Make a decision and interpret the results of the test.
Because 9.87 <u>does</u> fall in the critical region, we <u>reject the null hypothesis</u> of independence and conclude that approval of the President and political party are dependent.

15.

| (F$_o$) | Gender | | |
|---|---|---|---|
| Political Party Affiliation | Liberal | Conservative | Total |
| Republican | 15.3 | 12.7 | 28 |
| Democrat | 16.9 | 14.1 | 31 |
| Independent | 9.8 | 8.2 | 18 |
| Total | 42 | 35 | 77 |

$$\chi^2 (obtained) = \frac{(3-15.3)^2}{15.3} + \frac{(25-12.7)^2}{12.7} + \frac{(29-16.9)^2}{16.9} + \frac{(2-14.1)^2}{14.1}$$
$$+ \frac{(10-9.8)^2}{9.8} + \frac{(8-8.2)^2}{8.2} = 40.849$$

Step 1: Make assumptions and meet test requirements.
We were instructed to assume that all test assumptions/requirements (independent random samples and nominal level of measurement) have been met.

Step 2: State the null (and research) hypotheses.
H$_o$: The two variables are independent ($\chi^2$=0)
(H$_1$: The two variables are dependent)

Step 3: Select the sampling distribution and establish the critical region.
For α=0.05 at df=(r-1)(c-1)=(3-1)(2-1)=2, $\chi^2$(critical) equals 5.991.

Step 4: Compute the test statistic (see calculations above).
$\chi^2$(obtained)=40.849

Step 5: Make a decision and interpret the results of the test.
Because 40.849 <u>does</u> fall in the critical region, we <u>reject the null hypothesis</u> of independence and conclude that political party and political ideology are dependent.

16.

| (F$_o$) | Gender | | |
| Legalization of Marijuana | Male | Female | Total |
| --- | --- | --- | --- |
| Approve | 7.7 | 8.3 | 16 |
| Disapprove | 6.7 | 7.3 | 14 |
| No Opinion | 9.6 | 10.4 | 20 |
| Total | 24 | 26 | 50 |

$$\chi^2(obtained) = \frac{(10-7.7)^2}{7.7} + \frac{(6-8.3)^2}{8.3} + \frac{(5-6.7)^2}{6.7} + \frac{(9-7.3)^2}{7.3}$$

$$+ \frac{(9-9.6)^2}{9.6} + \frac{(11-10.4)^2}{10.4} = 2.229$$

Step 1: Make assumptions and meet test requirements.
We were instructed to assume that all test assumptions/requirements (independent random samples and nominal level of measurement) have been met.

Step 2: State the null (and research) hypotheses.
H$_o$: The two variables are independent ($\chi^2$=0)
(H$_1$: The two variables are dependent)

Step 3: Select the sampling distribution and establish the critical region.
For $\alpha$=0.05 at df=(r-1)(c-1)=(3-1)(2-1)=2, $\chi^2$(critical) equals 5.991.

Step 4: Compute the test statistic (see calculations above).
$\chi^2$(obtained)=2.229

Step 5: Make a decision and interpret the results of the test.
Because 2.229 does not fall in the critical region, we fail to reject the null hypothesis of independence and conclude that attitudes toward the legalization of marijuana and gender are dependent.

17.

| (F$_o$) | Place of Residence | | |
| Approval of Water Plant | Town | Farm | Total |
| --- | --- | --- | --- |
| Yes | 37.9 | 41.1 | 79 |
| No | 32.1 | 34.9 | 67 |
| Total | 70 | 76 | 146 |

$$\chi^2(obtained) = \frac{(33-37.9)^2}{37.9} + \frac{(46-41.1)^2}{41.1} + \frac{(37-32.1)^2}{32.1} + \frac{(30-34.9)^2}{34.9} = 2.65$$

Step 1: Make assumptions and meet test requirements.
We were instructed to assume that all test assumptions/requirements (independent random samples and nominal level of measurement) have been met.

Step 2: State the null (and research) hypotheses.
$H_o$: The two variables are independent ($\chi^2=0$)
($H_1$: The two variables are dependent)

Step 3: Select the sampling distribution and establish the critical region.
For $\alpha=0.05$ at df=(r-1)(c-1)=(2-1)(2-1)=1, $\chi^2$(critical) equals 3.841.

Step 4: Compute the test statistic (see calculations above).
$\chi^2$(obtained)=2.65

Step 5: Make a decision and interpret the results of the test.
Because 2.65 does <u>not</u> fall in the critical region, we <u>fail to reject the null hypothesis</u> of independence and conclude that approval of the water plant and place of residence are dependent.

## SPSS or MicroCase WORK PROBLEMS

If you had problems with any of these questions, be sure to re-read the appropriate sections at the end of chapter 10 in your textbook.

1. The following cross-tab was requested.

**Crosstabulation**

| | | | Sex | | Total |
|---|---|---|---|---|---|
| | | | MALE | FEMALE | |
| Abany | YES | Count | 117 | 143 | 260 |
| | | Column % | 43.8% | 37.9% | 40.4% |
| | NO | Count | 150 | 234 | 384 |
| | | Column % | 56.2% | 62.1% | 59.6% |
| Total | | Count | 267 | 377 | 644 |

**Chi-Square Tests**

| | Value | df | Asymp. Sig. (2-sided) |
|---|---|---|---|
| Pearson Chi-Square | 2.252(a) | 1 | .133 |
| N of Valid Cases | 644 | | |

a  0 cells (.0%) have expected count less than 5. The minimum expected count is 107.80.

Reading down the columns and comparing across rows, we see that 43.8% of the males favor abortion, whereas 37.9% of the females do so. And, 56.2% of males oppose abortion, while 62.1% of women do so. The significance value in the table (sig = 0.133) is greater than 0.05; therefore, we fail to reject the null hypothesis of independence. The evidence suggests that gender and abortion attitudes are independent.

2.    Using the data below, we see that 74.6% of the men and 84.8% of women say that extramarital sex is always wrong. 16.9% of men and 8.6% of women say that it is almost always wrong. 7.7% of men and 4.5% of women say that it is sometimes wrong, and .7% of men and 2.1% of women say that it is not wrong at all. In general, females were slightly more likely to think that extramarital sex is always wrong. Looking at the chi square output, below, we see that gender and attitudes toward extramarital sex are dependent (sig = 0.001, so we reject the null hypothesis of independence). The evidence suggests that men and women hold different views on extramarital sex.

**Crosstabulation**

| | | | Sex | | Total |
|---|---|---|---|---|---|
| | | | MALE | FEMALE | |
| Xmarsex | Always wrong | Count | 203 | 324 | 527 |
| | | Column % | 74.6% | 84.8% | 80.6% |
| | Almst always | Count | 46 | 33 | 79 |
| | wrong | Column % | 16.9% | 8.6% | 12.1% |
| | Sometimes | Count | 21 | 17 | 38 |
| | wrong | Column % | 7.7% | 4.5% | 5.8% |
| | Not wrong at | Count | 2 | 8 | 10 |
| | all | Column % | .7% | 2.1% | 1.5% |
| Total | | Count | 272 | 382 | 654 |

**Chi-Square Tests**

| | Value | df | Asymp. Sig. (2-sided) |
|---|---|---|---|
| Pearson Chi-Square | 15.890 (a) | 3 | .001 |
| N of Valid Cases | 654 | | |

a  1 cells (12.5%) have expected count less than 5. The minimum expected count is 4.16.

3.    Looking at the cross-tabs we see that more married people consider themselves "very happy" (45.5%) than any other category. Those who are widowed are more likely to consider themselves "not too happy" (20.8%) than any other category (although the divorced and separated respondents are close, with 204%). Those who are never married are most likely to think of themselves as "pretty happy" (60.3%). Looking at the chi square output, below, we see that these two variables, marital status and happiness, are dependent (sig = 0.000, we reject the null hypothesis of independence). The evidence suggests marital status and happiness are significantly related. (See next page.)

**Crosstabulation**

| | | | Marital | | | | Total |
|---|---|---|---|---|---|---|---|
| | | | MARRIED | WIDOWED | DIVORCED-SEPARATED | NEVER MARRIED | |
| Happy | VERY HAPPY | Count | 216 | 27 | 37 | 49 | 329 |
| | | Column % | 45.5% | 28.1% | 20.4% | 19.8% | 32.9% |
| | PRETTY HAPPY | Count | 221 | 49 | 107 | 149 | 526 |
| | | Column % | 46.5% | 51.0% | 59.1% | 60.3% | 52.7% |
| | NOT TOO HAPPY | Count | 38 | 20 | 37 | 49 | 144 |
| | | Column % | 8.0% | 20.8% | 20.4% | 19.8% | 14.4% |
| Total | | Count | 475 | 96 | 181 | 247 | 999 |

**Chi-Square Tests**

| | Value | df | Asymp. Sig. (2-sided) |
|---|---|---|---|
| Pearson Chi-Square | 78.328 (a) | 6 | .000 |
| N of Valid Cases | 999 | | |

a  0 cells (.0%) have expected count less than 5. The minimum expected count is 13.84.

4.   Using the data below, we see that 89.5% of the men and 92.3% of women favor sex education.  10.5% of men and 7.7% of women oppose it.  In general, females were slightly more likely to favor sex education than men (and men are more likely to oppose it than women).  Looking at the chi square output, below, we see that gender and attitudes toward sex education are independent (sig = 0.201, so we fail to reject the null hypothesis of independence).  The evidence suggests that men and women hold different views on sex education.

**Crosstabulation**

| | | | Sex | | Total |
|---|---|---|---|---|---|
| | | | MALE | FEMALE | |
| Sexeduc | Favor | Count | 255 | 349 | 604 |
| | | Column % | 89.5% | 92.3% | 91.1% |
| | Oppose | Count | 30 | 29 | 59 |
| | | Column % | 10.5% | 7.7% | 8.9% |
| Total | | Count | 285 | 378 | 663 |

**Chi-Square Tests**

| | Value | df | Asymp. Sig. (2-sided) |
|---|---|---|---|
| Pearson Chi-Square | 1.633 (a) | 1 | .201 |
| N of Valid Cases | 663 | | |

a  0 cells (.0%) have expected count less than 5. The minimum expected count is 25.36.

5.   Using the data below, we see that 23.8% of the men and 29.3% of women say that premarital sex is always wrong. 4.3% of men and 9.2% of women say that it is almost always wrong. 19.9% of men and 21.5% of women say that it is sometimes wrong, and 52.1% of men and 39.9% of women say that it is not wrong at all. In general, females were slightly more likely to think that premarital sex is always wrong. Looking at the chi square output, below, we see that gender and attitudes toward premarital sex are dependent (sig = 0.005, so we reject the null hypothesis of independence). The evidence suggests that men and women hold different views on premarital sex.

**Crosstabulation**

| | | | Sex | | Total |
|---|---|---|---|---|---|
| | | | MALE | FEMALE | |
| Premarsx | Always wrong | Count | 67 | 108 | 175 |
| | | Column % | 23.8% | 29.3% | 26.9% |
| | Almst always | Count | 12 | 34 | 46 |
| | wrong | Column % | 4.3% | 9.2% | 7.1% |
| | Sometimes | Count | 56 | 79 | 135 |
| | wrong | Column % | 19.9% | 21.5% | 20.8% |
| | Not wrong at | Count | 147 | 147 | 294 |
| | all | Column % | 52.1% | 39.9% | 45.2% |
| Total | | Count | 282 | 368 | 650 |

**Chi-Square Tests**

| | Value | df | Asymp. Sig. (2-sided) |
|---|---|---|---|
| Pearson Chi-Square | 12.893 (a) | 3 | .005 |
| N of Valid Cases | 650 | | |

a  0 cells (.0%) have expected count less than 5. The minimum expected count is 19.96.

# CHAPTER TWELVE
## Introduction to Bivariate Association and Measures of Association for Variables Measured at the Nominal Level

*Learning Objectives:* By the end of this chapter, you will be able to:

1. Explain how we can use measures of association to describe and analyze the importance of relationships versus their statistical significance.
2. Define "association" in the context of bivariate tables and in terms of changing conditional distributions.
3. List and explain the three characteristics of a bivariate relationship: existence, strength, and pattern or direction.
4. Investigate a bivariate association by properly calculating percentages for a bivariate table and interpret results.
5. Compute and interpret measures of association for variables measured at the nominal level.

## KEY TERMS

- ✓ Association (p. 283)
- ✓ Column Percentages (p. 285)
- ✓ Conditional Distribution of Y (p. 284)
- ✓ Cramer's V (p. 292)
- ✓ Dependent Variable (p. 283)
- ✓ Independent Variable (p. 283)
- ✓ Lambda (p. 295)
- ✓ Maximum Difference (p. 287)
- ✓ Measures of Association (p. 282)
- ✓ Negative Association (p. 288)
- ✓ Phi ($\varphi$) (p. 291)
- ✓ Positive Association (p. 288)
- ✓ Proportional Reduction in Error (PRE) (p. 295)
- ✓ X (p. 283)
- ✓ Y (p. 283)

## CHAPTER SUMMARY

This chapter introduces us to the concept of bivariate (meaning *two* variables) association or the relationship of one variable with another. Two variables are considered to have an association when the distribution of one changes under certain conditions of the other – or when one variable has an *effect* on the other variable. For example, being a Democrat (one condition of independent variable) is associated with (or affects) the likelihood of being pro-choice or pro-life. Thus, we can say that political party affiliation and attitudes toward abortion are associated.

# CHAPTER OUTLINE

## 12.1 Statistical Significance and Theoretical Importance.

## 12.2 Association Between Variables and Bivariate Tables.

## 12.3 Three Characteristics of Bivariate Associations.

- Measures of association are interested in the following three questions: (1) Does an association exist between two variables? (2) How strong is the association? and (3) What is the direction or pattern of association?

- *Does an association exist?* To determine if an association exists, we look at the changes in the distribution of the dependent variable (Y) across all categories of the independent variable (X). In a bivariate table the dependent variable is arrayed in the rows, and the independent variable is arrayed in the columns. We read this table by reading down the columns (in the direction of the independent variable), but comparing across the rows (in the direction of the dependent variable). This is called the conditional distribution of Y, or the distribution of the dependent variable given a certain condition of the independent variable (e.g., being either Republican or Democrat).

- *What is the direction or pattern of association?* The direction or pattern of the relationship tells us exactly *how* the two variables are related. We talk about the direction of the relationship for variables at the ordinal or interval-ratio level only. Variables measured at the nominal level cannot be described in terms of direction because there is no ordering of the values in nominal-level variables; however, we can talk of the pattern of the relationship between nominal variables. When we discuss the direction of a relationship, we talk about positive or negative relationships. A positive relationship is one in which the two variables change in the same direction, that is, as education increases, income increases. Similarly, the statement "people with lower education have lower incomes" also reflects a positive relationship because the direction of the two variables is the same. A negative association is one in which the relationship between the two variables is going in the opposite direction, that is, as education increases, the number of children one has decreases.

- *How strong is the association?* For a "quick and easy" method for assessing the strength of the relationship, the maximum difference technique can be used. The maximum difference technique looks at the differences between column percentages to determine if the relationship is <u>weak</u> (the maximum difference in column percents in the table is between 0 and 10 percentage points), <u>moderate</u> (between 10 and 30 percentage points), or <u>strong</u> (more than 30 percentage points). Although this technique will give you a *sense* of the relationship between the two variables, the true strength of the relationship should be measured by a statistic (such as phi, Cramer's V, and lambda, which are discussed below). The specific statistic to be used is dictated by the level of measurement of the variables in question.

## 12.4 Introduction to Measures of Association.

## 12.5 Measures of Association for Variables Measured at the Nominal Level: Chi Square Based Measures.

- Three measures of association for nominal data are introduced in chapter 11, phi (M), Cramer's V, and lambda (8). The first two measures, phi and Cramer's V, are chi square based measures, which means they are calculated using the chi square statistic. Phi is used for 2x2 tables, and V is used for tables larger than 2x2. Both are good measures, but interpretation is a bit tough. With both of these measures there is no real statistical interpretation, only a generalization. Thus, a phi of 0.33 seems a bit weak, and we know it is weaker than a phi of 0.54, but exactly how much weaker is unknown. (Your textbook recommends using the "maximum difference" scale to interpret the values of phi and V: between 0 and .1 is weak, .11 to .3 is moderate, and greater than .3 is strong.) Therefore, it might be better to use another type of measure, like lambda.

## 12.6 Lambda: A Proportional Reduction in Error Measure of Association for Nominal Level Variables.

- Lambda is a PRE measure, which stands for Proportional Reduction in Error. PRE measures like lambda indicate the amount of error we reduce in predicting the dependent variable with knowledge of the independent variable (in comparison to having no knowledge of the independent variable). In your textbook, Healey introduces a nice example of the logic. Basically it works like this: let's say you have to predict the height of every person walking through a door as either tall (over 5'9") or short (under 5'9"), with no information about the person. You have a 50/50 chance of correctly guessing the person's height. What if you are told that the next person to walk through that door will be female. Logically, we can say that, on average, women are shorter than 5'9" (average female height is 5'4"). If we predict that she will be "short," there is a good possibility that we will be correct (better than 50/50). Therefore, knowledge of the person's gender will likely reduce the number of errors we make in predicting height.

- Try to keep this example in mind whenever you are working with PRE measures (other PRE measures will be discussed in later chapters). Remember that you are trying to reduce the number of errors you make in predicting the dependent variable by having knowledge of the independent variable.

- PRE measures range from zero to 1.0, with zero indicating no improvement whatsoever in our prediction of the dependent variable with the independent variable, and 1.0 indicating that the independent variable perfectly predicts the dependent variable. Lambda can also be interpreted as a percentage. If we have a lambda of 0.55, we can convert that to a percentage and state that we have reduced the amount of error in predicting the dependent variable by 55% with knowledge of the independent variable.

- One problem with lambda that you might encounter is that when one row marginal is exceptionally large, lambda might be zero. This does not necessarily mean that there is no association between these two variables, only that the large row marginal has made lambda act strangely. If this is the case, use a chi square based measure of association (either phi or V) to assess the strength of the relationship.

- A second problem with lambda (but not with phi or V) is that it is an asymmetric measure. This means that the value of lambda depends on which variable is treated as the dependent variable. Thus, the value of lambda when predicting height with knowledge of gender will be different than the value of lambda when predicting gender with knowledge of height. You should always take care when determining which variable is the dependent and which is the independent.

## MULTIPLE-CHOICE QUESTIONS

1. For an association to exist, the _____ for one variable changes for the _____ of another variable.
   a. distribution, categories
   b. categories, distribution
   c. distribution, distribution
   d. cases, means

   *Learning Objective: 2*

2. The dependent variable is thought to be _____, and the independent variable is thought to be _____.
   a. the cause, the effect
   b. the effect, the cause
   c. positive, negative
   d. X, Y

   *Learning Objective: 1*

3. X is used to identify the _____ variable, and Y is used to identify the _____ variable.
   a. independent, dependent
   b. dependent, independent
   c. effect, cause
   d. nominal, interval-ratio

   *Learning Objective: 1*

4.  A negative association is one in which the variables vary in _____ directions, and a positive association is one in which the variables vary in _____ directions.
    a.  the same, opposite
    b.  opposite, the same
    c.  negative, the same
    d.  opposite, positive

    *Learning Objective: 2*


5.  What three questions can we ask (and answer) about a bivariate relationship?
    a.  Does an association exist? What is the strength? What should be the logical next step?
    b.  Does the independent variable cause the effect? What is the direction of the relationship? Does an association exist?
    c.  Does an association exist? Is there a third variable? What is the direction of the association?
    d.  Does an association exist? What is the strength of the association? What is the direction of the association?

    *Learning Objective: 3*


6.  Percentages are calculated in the direction of the _____ variable.
    a.  independent
    b.  dependent
    c.  bivariate
    d.  nominal

    *Learning Objective: 4*


7.  When creating a bivariate table, the _____ variable should be placed along the top of the table (in the columns), and the _____ variable should be placed along the side of the table (in the rows).
    a.  nominal, ordinal
    b.  ordinal, nominal
    c.  dependent, independent
    d.  independent, dependent

    *Learning Objective: 4*

8.    If one variable increases and the other variable decreases, the association is said to be
      _____.
      a.  positive
      b.  different
      c.  negative
      d.  similar

*Learning Objective: 2*


9.    The chi-square based measures of association are _____ and _____.
      a.  lambda, phi
      b.  lambda, Cramer's V
      c.  phi and V
      d.  phi and chi

*Learning Objective: 5*


10.   The phi measure is used on _____ tables, and the Cramer's V measure is used on
      _____ tables.
      a.  nominal, ordinal
      b.  2x2, larger than 2x2
      c.  larger than 2x2, 2x2
      d.  multivariate, bivariate

*Learning Objective: 5*


11.   The PRE measure of association for nominal data is _____.
      a.  phi
      b.  Cramer's V
      c.  lambda
      d.  chi square

*Learning Objective: 5*


12.   PRE means that we reduce the error in predicting the _____ with knowledge of the
      _____.
      a.  PRE measure, independent variable
      b.  independent variable, dependent variable
      c.  PRE measure, chi square
      d.  dependent variable, independent variable

*Learning Objective: 5*

13. Lambda might equal zero when _____.
   a. one row marginal is exceptionally large.
   b. there is an association between the two variables.
   c. there is a large number of missing data.
   d. the dependent variable is ordinal.

   *Learning Objective: 5*

14. Lambda is an asymmetric measure, meaning it _____.
   a. cannot be used with nominal data.
   b. will change value depending on which variable is dependent and which is independent.
   c. is not a PRE measure.
   d. is not appropriate to use when there are two or more categories.

   *Learning Objective: 5*

15. Lambda ranges from _____.
   a. - 1 to + 1
   b. -1 to 0
   c. 0 to infinity
   d. 0 to +1

   *Learning Objective: 5*

16. If two variables share an association, then there is sufficient evidence to conclude that they also share a causal relationship.
   a. true
   b. false

   *Learning Objective: 2*

17. The conditional distributions of Y _____.
   a. display the distribution of scores on the dependent variable for each score on the independent variable
   b. display the distribution of scores on the independent variable for each score on the dependent variable
   c. are used to prove an association between the dependent and independent variables
   d. are used when the dependent variable is arrayed in the columns of a bivariate table

   *Learning Objective: 2*

18. In order to determine the pattern or direction of an association between two variables, compare _____.
   a. row percents
   b. $\chi^2$(obtained) values
   c. column percents
   d. lambda

   *Learning Objective: 4*


19. If two variables are unrelated, then their conditional distributions of Y _____.
   a. would be different from each other
   b. could not be calculated
   c. would all be located in one and only one cell
   d. would be the same

   *Learning Objective: 4*


20. A _____ exists between two variables if the value of the dependent variable is associated with one and only one value of the independent variable.
   a. perfect non-association
   b. perfect association
   c. negative association
   d. positive association

   *Learning Objective: 3*


## WORK PROBLEMS


1. If a researcher is interested in the relationship between age and education, which variable would be independent and which would be dependent?

   *Learning Objective: 2*


2. This same researcher is interested in education and age first married, hypothesizing that more educated people marry later. Which is the independent variable and which is the dependent variable?

   *Learning Objective: 2*

3. Quality of life and health status are thought to be associated. Make a hypothesis and indicate which variable is independent and which is dependent.

   *Learning Objective: 2*

4. Another researcher is interested in testing the hypothesis that increased religiosity decreases support for same-sex marriage. Which is the independent variable and which is the dependent variable?

   *Learning Objective: 2*

5. Let's say that the researcher in the previous problem collected the following data. Looking only at the conditional distributions of Y, does there appear to be an association?

   | ($F_o$) | Religiosity | | | |
   | Support for Same-Sex Marriage | Low | Moderate | High | Total |
   |---|---|---|---|---|
   | No | 15 | 40 | 55 | 110 |
   | Yes | 50 | 30 | 10 | 90 |
   | Total | 65 | 70 | 65 | 200 |

   *Learning Objective: 2*

6. Using the data from the previous problem, calculate column percents to determine the pattern/direction of the association. What is the strength of the association?
   $\chi2$(obtained)=49.92

   *Learning Objective: 4 & 5*

7. What is the pattern/direction of association between these two variables?

   | ($F_o$) | Gender | | |
   | Marital Satisfaction | Male | Female | Total |
   |---|---|---|---|
   | Very Happy | 30 | 45 | 75 |
   | Moderately Happy | 50 | 45 | 95 |
   | Unhappy | 20 | 10 | 30 |
   | Total | 100 | 100 | 200 |

   *Learning Objective: 4*

8. Using the information from problem 7, calculate Cramer's V and lambda. What do they tell us about the strength of the relationship between gender and marital satisfaction? $\Pi^2$(obtained)=6.66

*Learning Objective: 5*

9. What is the pattern/direction of association between these two variables?

| ($F_o$) | Gender | | |
|---|---|---|---|
| Attitude Toward Gun Control | Male | Female | Total |
| Favor | 45 | 42 | 87 |
| Oppose | 39 | 49 | 88 |
| Total | 84 | 91 | 175 |

*Learning Objective: 4*

10. Using the information from problem 9, calculate phi and lambda. What do they tell us about the strength of the relationship between gender and attitudes toward gun control? $\Pi^2$(obtained)=0.96

*Learning Objective: 5*

11. A researcher is interested in knowledge of birth control between men and women. What is the pattern/direction of association between these two variables?

| ($F_o$) | Gender | | |
|---|---|---|---|
| Knowledge of Birth Control | Male | Female | Total |
| High | 175 | 430 | 605 |
| Moderate | 512 | 220 | 732 |
| Low | 215 | 160 | 375 |
| Total | 902 | 810 | 1710 |

*Learning Objective: 4*

12. Using the information from problem 11, calculate and interpret the best strength measure for this relationship. $\Pi^2$(obtained)=227.7

*Learning Objective: 5*

13. In a study of attitudes toward legalization of marijuana, the following data were collected. Are males more likely to favor legalization than females?

| $(F_o)$ | Gender | | |
|---|---|---|---|
| Legalization of Marijuana | Male | Female | Total |
| Approve | 10 | 6 | 16 |
| Disapprove | 5 | 9 | 14 |
| No Opinion | 9 | 11 | 20 |
| Total | 24 | 26 | 50 |

*Learning Objective: 4*

14. Using the information from problem 13, calculate and interpret all appropriate measures of strength for this relationship. $\Pi^2$(obtained)=2.23

*Learning Objective: 5*

15. Using the data below, answer all three questions posed in this chapter. Does an association exist? How strong is the association? What is the pattern/direction of association? $\Pi^2$(obtained)=9.87

| $(F_o)$ | Political Party Preference | | |
|---|---|---|---|
| Approval of President | Republican | Democrat | Total |
| Yes | 14 | 29 | 43 |
| No | 24 | 11 | 35 |
| Total | 38 | 40 | 78 |

*Learning Objective: 4 & 5*

## SPSS or MicroCase WORK PROBLEMS

The following work problems can be completed using either SPSS or MicroCase. The solutions will be the same regardless of which one you use. Your instructor may have a preference, so be sure to check with him/her prior to beginning these problems.

1. Using the GSS 2006 dataset, discuss the relationship between abortion (*abany*) and gender (*sex*). Be sure to look at the conditional distributions of Y in terms of percentages. Answer all three questions: Is there an association? How strong is the association? What is the pattern/direction of association?

2.    Using the GSS 2006 dataset, discuss the relationship between attitudes toward extramarital sex (*xmarsex*) and gender (*sex*). Be sure to look at the conditional distributions of Y in terms of percentages. Answer all three questions: Is there an association? How strong is the association? What is the pattern/direction of association?

3.    Using the GSS 2006 dataset, discuss the relationship between happiness (*happy*) and marital status (*marital*). Be sure to look at the conditional distributions of Y in terms of percentages. Answer all three questions: Is there an association? How strong is the association? What is the pattern/direction of association?

4.    Using the GSS 2006 dataset, discuss the relationship between attitudes sex education (*sexeduc*) and gender (*sex*). Be sure to look at the conditional distributions of Y in terms of percentages. Answer all three questions: Is there an association? How strong is the association? What is the pattern/direction of association?

5.    Which group is more likely to believe that sex before marriage (*premarsx*) is wrong – males or females? Using the GSS 2006 dataset, analyze this relationship. Be sure to look at the conditional distributions of Y in terms of percentages. Answer all three questions: Is there an association? How strong is the association? What is the pattern/direction of association?

**ANSWER KEY**

**MULTIPLE-CHOICE QUESTIONS**

| | | | | | | |
|---|---|---|---|---|---|---|
| 1. | a | (p. 282-283) | 11. | c | (p. 295) |
| 2. | b | (p. 283) | 12. | c | (p. 295) |
| 3. | a | (p. 283) | 13. | c | (p. 298) |
| 4. | b | (p. 288) | 14. | a | (p. 298) |
| 5. | d | (p. 285) | 15. | b | (p. 297) |
| 6. | a | (p. 285) | 16. | b | (p. 283) |
| 7. | d | (p. 283) | 17. | a | (p. 284) |
| 8. | c | (p. 288) | 18. | c | (p. 285) |
| 9. | c | (p. 290-292) | 19. | d | (p. 284) |
| 10. | b | (p. 291) | 20. | b | (p. 286) |

## WORK PROBLEMS

1.  Because age predicts when you begin school, and the older you get the more schooling you attain, age is the independent variable. Another way to think about it is: age increases the amount of schooling one has, education does not increase someone's age.

2.  As stated in the hypothesis, education is the independent (or "causal") variable, which is hypothesized to have a positive effect on age first married, which is the dependent variable.

3.  Hypothesis: *Health status will have a positive effect on quality of life.* As one's health decreases, so too will his/her quality of life. Therefore, health status is the independent variable, and quality of life is the dependent variable.

4.  The researcher seeks to test whether higher levels of religiosity lead to lower levels of support for same-sex marriage. Because religiosity is taking the causal role (and support for same-sex marriage is the outcome), religiosity is the independent variable and same-sex marriage support is the dependent variable.

5.  Because the conditional distributions of Y are not all the same, we can conclude that there is an association between religiosity and support for same-sex marriage.

6.

| ($F_o$) | Religiosity | | | |
| Support for Same-Sex Marriage | Low | Moderate | High | Total |
| --- | --- | --- | --- | --- |
| No | 15 (23%) | 40 (57%) | 55 (85%) | 110 |
| Yes | 50 (77%) | 30 (43%) | 10 (15%) | 90 |
| Total | 65 | 70 | 65 | 200 |

The column percents indicate that support for same-sex marriage is associated with lower levels of religiosity, while opposition to same-sex marriage is associated with higher religiosity.

We were given the $\Pi^2$(obtained) value of 49.92. We can use this value to calculate Cramer's V. (Phi is not an appropriate measure of association for this table because it is larger than 2x2.)

$$V = \sqrt{\frac{\chi^2}{N(\min r - 1, c - 1)}} = \sqrt{\frac{49.92}{200(2-1)}} = 0.50$$

$$E_1 = N - (largest\ row\ total) = 200 - 110 = 90$$

$$E_2 = (65 - 50) + (70 - 40) + (65 - 55) = 55$$

$$\lambda = \frac{E_1 - E_2}{E_1} = \frac{90 - 55}{90} = 0.39$$

Therefore, there is a strong association between these two variables (V=0.50). According to lambda, we reduce our error in predicting the dependent variable by 39 percent when we have knowledge of religiosity.

7.

| ($F_o$ and Column %) | Gender | | |
| --- | --- | --- | --- |
| Marital Satisfaction | Male | Female | Total |
| Very Happy | 30 (30%) | 45 (45%) | 75 |
| Moderately Happy | 50 (50%) | 45 (45%) | 95 |
| Unhappy | 20 (20%) | 10 (10%) | 30 |
| Total | 100 | 100 | 200 |

Because the column percents are fairly close, there is insufficient evidence to suggest an association between these two variables. It appears that more women state that they are very happy with their marriage (45% compared to 30% for males), yet the moderately happy category is almost even across both groups (50% for males, 45% for females) and males are only slightly more likely than females to be unhappy (20% for males, 10% for females).

8. We were given the $\Pi^2$(obtained) value of 6.66. We can use this value to calculate Cramer's V. (Phi is not an appropriate measure of association for this table because it is larger than 2x2.)

$$V = \sqrt{\frac{\chi^2}{N(\min r - 1, c - 1)}} = \sqrt{\frac{6.66}{200(2-1)}} = 0.18$$

$$E_1 = N - (largest\ row\ total) = 200 - 95 = 105$$

$$E_2 = \sum_{column} (column\ marginal - largest\ cell\ frequency) = (100 - 50) + (100 - 45) = 105$$

$$\lambda = \frac{E_1 - E_2}{E_1} = \frac{105 - 105}{105} = 0$$

Therefore, there is a moderate association between these two variables (V=0.18). We cannot talk about the PRE measure because lambda was equal to zero. (Lambda can be zero when on of the row marginals is larger than the others.) The maximum difference in the table is .15 (between very happy males and very happy females); therefore, according to the interpretation scale in your textbook, this is a weakly moderate association. You may recall that the chi square hypothesis test in chapter 11 (work problem 6) found gender and marital satisfaction to be dependent. Thus, it appears that while the two variables are dependent, their association is limited.

9.

| ($F_o$ and Column %) | Gender | | |
| Attitude Toward Gun Control | Male | Female | Total |
| --- | --- | --- | --- |
| Favor | 45 (54%) | 42 (46%) | 87 |
| Oppose | 39 (46%) | 49 (49%) | 88 |
| Total | 84 | 91 | 175 |

Again, the column percents are close, so there is insufficient evidence to suggest an association between these two variables. Men are slightly more likely to favor gun control (54% for males, 46% for females), and conversely, women are slightly more likely to oppose it (46% for males, 49% for females).

10. We were given the $\Pi^2$(obtained) value of 0.96. We can use this value to calculate phi. (V is not an appropriate measure of association for this table because it is 2x2.)

$$\Phi = \sqrt{\frac{\chi^2}{N}} = \sqrt{\frac{0.96}{175}} = 0.07$$

$$E_1 = N - \text{(largest row total)} = 175 - 88 = 87$$

$$E_2 = \sum_{column} \left(\text{column marginal} - \text{largest cell frequency}\right) = (84 - 45) + (91 - 49) = 81$$

$$\lambda = \frac{E_1 - E_2}{E_1} = \frac{87 - 81}{87} = 0.07$$

According to phi, there is a weak association between these two variables. In addition, we reduce the amount of error in predicting attitudes toward gun control by 7% when we have knowledge of gender. Moreover, the maximum difference in column percents is 0.08 (between males who favor and females who favor), which indicates a weak relationship. Therefore, all strength measures indicate a weak association between gender and attitudes toward gun control. You may recall that the chi square hypothesis test in chapter 11 (work problem 7) also found gender and gun control attitudes to be independent. Taken together, all measures indicate a weak to no association.

11.

| (F$_0$ and Column %) | Gender | | |
| Knowledge of Birth Control | Male | Female | Total |
| --- | --- | --- | --- |
| High | 175 (19%) | 430 (53%) | 605 |
| Moderate | 512 (57%) | 220 (27%) | 732 |
| Low | 215 (24%) | 160 (20%) | 375 |
| Total | 902 | 810 | 1710 |

These two variables appear to be associated, because, when comparing the conditional distributions of Y, women appear to be more knowledgeable than men. For example, women are more likely than men to have high knowledge of birth control (19% for males, 53% for females). Men, however, are more likely than women to have low (24% for males, 20% for females) or moderate (57% for males, 27% for females) knowledge.

12. As a PRE measure, lambda is considered the best strength measure for nominal variables.

$$E_1 = N - (\text{largest row total}) = 1710 - 732 = 980$$

$$E_2 = \sum_{column} (\text{column marginal} - \text{largest cell frequency}) = (902 - 512) + (810 - 430) = 770$$

$$\lambda = \frac{E_1 - E_2}{E_1} = \frac{980 - 770}{980} = 0.21$$

According to lambda, we reduce the amount of error in predicting knowledge of birth control by 21% when we have knowledge of gender. In addition, the maximum difference in column percents is 0.34 (between males with high knowledge and females with high knowledge), which indicates a strong relationship. You may recall that the chi square hypothesis test in chapter 11 (work problem 8) also found gender and knowledge of birth control to be dependent. Taken together, all measures indicate a moderate to strong association.

13.

| (F$_0$ and Column %) | Gender | | |
| Legalization of Marijuana | Male | Female | Total |
| --- | --- | --- | --- |
| Approve | 10 (42%) | 6 (23%) | 16 |
| Disapprove | 5 (21%) | 9 (35%) | 14 |
| No Opinion | 9 (37%) | 11 (42%) | 20 |
| Total | 24 | 26 | 50 |

There appears to be a weak association between gender and attitudes toward legalizing marijuana. We can conclude that men are more likely than women to favor legalization.

14. We were given the $\Pi^2$(obtained) value of 2.23. We can use this value to calculate Cramer's V. (Phi is not an appropriate measure of association for this table because it is larger than 2x2.)

$$V = \sqrt{\frac{\chi^2}{N(\min r-1, c-1)}} = \sqrt{\frac{2.23}{175(2-1)}} = 0.15$$

$E_1 = N - (\text{largest row total}) = 50 - 20 = 30$

$E_2 = \sum_{column}(\text{column marginal} - \text{largest cell frequency}) = (24-10)+(26-11) = 29$

$\lambda = \dfrac{E_1 - E_2}{E_1} = \dfrac{30-29}{30} = 0.03$

According to lambda, we reduce the amount of error in predicting attitudes toward the legalization of marijuana by 3% when we have knowledge of gender. Cramer's V suggests a slightly stronger relationship than lambda does. Cramer's V suggests a weakly moderate relationship between the two variables. In addition, the maximum difference in column percents is 0.19 (between males who approve and females who approve), which indicates a moderate relationship. You may recall that the chi square hypothesis test in chapter 11 (work problem 16) found gender and attitudes toward the legalization of marijuana to be independent. Taken together, all measures indicate a weak to moderate association.

15.

| (F$_o$ and Column %) | Political Party Preference | | |
|---|---|---|---|
| Approval of President | Republican | Democrat | Total |
| Yes | 14 (36.8%) | 29 (72.5%) | 43 |
| No | 24 (63.2%) | 11 (27.5%) | 35 |
| Total | 38 | 40 | 78 |

We were given the $\Pi^2$(obtained) value of 9.87. We can use this value to calculate phi. (Cramer's V is not an appropriate measure of association for this table because it is 2x2.)

$$\Phi = \sqrt{\frac{\chi^2}{N}} = \sqrt{\frac{9.87}{78}} = 0.36$$

$E_1 = N - (\text{largest row total}) = 78 - 43 = 35$

$E_2 = \sum_{column}(\text{column marginal} - \text{largest cell frequency}) = (38-24)+(40-29) = 25$

$\lambda = \dfrac{E_1 - E_2}{E_1} = \dfrac{35-25}{35} = 0.29$

275

The column percents indicate that democrats are more likely to approve of the president (72.5% for democrats, 36.8% for republicans), and republicans are more likely to disapprove of the president (27.5% for democrats, 63.2% for republicans).

According to lambda, we reduce the amount of error in predicting approval of the President by 29% when we have knowledge of political party. Phi suggests a strong relationship between the two variables. In addition, the maximum difference in column percents is 0.357, which also indicates a strong relationship. You may recall that the chi square hypothesis test in chapter 11 (work problem 14) also found political party and approval of the President to be dependent. Taken together, all measures indicate a strong association.

## SPSS or MicroCase WORK PROBLEMS

If you had problems with any of these questions, be sure to re-read the appropriate sections at the end of the chapter in your textbook.

1. *Association:* The chi square hypothesis test indicates that gender and abortion attitudes are independent. (Because the significance level is greater than 0.05 we fail to reject the null hypothesis of independence.) The evidence suggests that these two variables are unrelated.

   *Pattern:* Starting in the column for males, 43.8% are in favor of abortion, compared to 37.9% of the females. 56.2% of the males are opposed to abortion, compared to 62.1% of the females. Thus, females are a bit more likely to oppose abortion than men.

   *Strength:* Lambda indicates that no improvements were made in predicting attitudes toward abortion given information on the respondent's gender (value =0.000). And phi equals .059, which also indicates a very weak relationship.

**Crosstabulation**

| | | | Sex | | Total |
| --- | --- | --- | --- | --- | --- |
| | | | MALE | FEMALE | |
| Abany | YES | Count | 117 | 143 | 260 |
| | | Column % | 43.8% | 37.9% | 40.4% |
| | NO | Count | 150 | 234 | 384 |
| | | Column % | 56.2% | 62.1% | 59.6% |
| Total | | Count | 267 | 377 | 644 |

**Chi-Square Tests**

| | Value | df | Asymp. Sig. (2-sided) |
| --- | --- | --- | --- |
| Pearson Chi-Square | 2.252(a) | 1 | .133 |
| N of Valid Cases | 644 | | |

a  0 cells (.0%) have expected count less than 5. The minimum expected count is 107.80.

**Directional Measures**

| | | Value | Asymp. Std. Error(a) | Approx. T | Approx. Sig. |
|---|---|---|---|---|---|
| Lambda | Symmetric | .000 | .000 | .(b) | .(b) |
| | Abany Dependent | .000 | .000 | .(b) | .(b) |
| | Sex Dependent | .000 | .000 | .(b) | .(b) |

a Not assuming the null hypothesis.
b Cannot be computed because the asymptotic standard error equals zero.

**Symmetric Measures**

| | Value | Approx. Sig. |
|---|---|---|
| Phi | .059 | .133 |
| Cramer's V | .059 | .133 |
| N of Valid Cases | 644 | |

a Not assuming the null hypothesis.
b Using the asymptotic standard error assuming the null hypothesis.

2. *Association:* Looking at the chi square output, below, we see that gender and attitudes toward extramarital sex are dependent (sig = 0.001, so we reject the null hypothesis of independence). The evidence suggests that men and women hold different views on extramarital sex.

*Pattern:* Using the data below, we see that 74.6% of the men and 84.8% of women say that extramarital sex is always wrong. 16.9% of men and 8.6% of women say that it is almost always wrong. 7.7% of men and 4.5% of women say that it is sometimes wrong, and .7% of men and 2.1% of women say that it is not wrong at all. In general, females were slightly more likely to think that extramarital sex is always wrong.

*Strength:* Lambda equals .000 (mostly because of the uneven column marginals, nearly everyone, 80.6%, thinks extramarital sex is always wrong), indicating a very weak relationship. Cramer's V indicates that gender and extramarital sex attitudes are weakly related (V=0.156).

**Crosstabulation**

| | | | Sex | | Total |
|---|---|---|---|---|---|
| | | | MALE | FEMALE | |
| Xmarsex | Always wrong | Count | 203 | 324 | 527 |
| | | Column % | 74.6% | 84.8% | 80.6% |
| | Almst always wrong | Count | 46 | 33 | 79 |
| | | Column % | 16.9% | 8.6% | 12.1% |
| | Sometimes wrong | Count | 21 | 17 | 38 |
| | | Column % | 7.7% | 4.5% | 5.8% |
| | Not wrong at all | Count | 2 | 8 | 10 |
| | | Column % | .7% | 2.1% | 1.5% |
| Total | | Count | 272 | 382 | 654 |

**Chi-Square Tests**

| | Value | df | Asymp. Sig. (2-sided) |
|---|---|---|---|
| Pearson Chi-Square | 15.890 (a) | 3 | .001 |
| N of Valid Cases | 654 | | |

a  1 cells (12.5%) have expected count less than 5. The minimum expected count is 4.16.

**Directional Measures**

| | | Value | Asymp. Std. Error(a) | Approx. T | Approx. Sig. |
|---|---|---|---|---|---|
| Lambda | Symmetric | .043 | .026 | 1.575 | .115 |
| | Xmarsex Dependent | .000 | .000 | .(b) | .(b) |
| | Sex Dependent | .036 | .039 | 1.575 | .115 |

a  Not assuming the null hypothesis.
b  Cannot be computed because the asymptotic standard error equals zero.

**Symmetric Measures**

| | Value | Approx. Sig. |
|---|---|---|
| Phi | .156 | .001 |
| Cramer's V | .156 | .001 |
| N of Valid Cases | 654 | |

a  Not assuming the null hypothesis.
b  Using the asymptotic standard error assuming the null hypothesis.

3.   *Association:*  Looking at the chi square output, below, we see that marital status and happiness are dependent (sig = 0.001, so we reject the null hypothesis of independence).

*Pattern:*  Looking at the cross-tabs we see that more married people consider themselves "very  happy" (45.5%) than any other category.  Those who are widowed are more likely to consider themselves "not too happy" (20.8%) than any other category (although the divorced and separated respondents are close, with 204%).  Those who are never married are most likely to think of themselves as "pretty happy" (60.3%).

*Strength:*  Lambda equals .000 again, indicating a weak to no relationship. Cramer's V indicates that a moderate relationship (V=0.198).

**Crosstabulation**

| | | | Marital | | | | Total |
|---|---|---|---|---|---|---|---|
| | | | MARRIED | WIDOWED | DIVORCED-SEPARATED | NEVER MARRIED | |
| Happy | VERY HAPPY | Count | 216 | 27 | 37 | 49 | 329 |
| | | Column % | 45.5% | 28.1% | 20.4% | 19.8% | 32.9% |
| | PRETTY HAPPY | Count | 221 | 49 | 107 | 149 | 526 |
| | | Column % | 46.5% | 51.0% | 59.1% | 60.3% | 52.7% |
| | NOT TOO HAPPY | Count | 38 | 20 | 37 | 49 | 144 |
| | | Column % | 8.0% | 20.8% | 20.4% | 19.8% | 14.4% |
| Total | | Count | 475 | 96 | 181 | 247 | 999 |

**Chi-Square Tests**

|  | Value | df | Asymp. Sig. (2-sided) |
|---|---|---|---|
| Pearson Chi-Square | 78.328 (a) | 6 | .000 |
| N of Valid Cases | 999 |  |  |

a 0 cells (.0%) have expected count less than 5. The minimum expected count is 13.84.

**Directional Measures**

|  |  | Value | Asymp. Std. Error(a) | Approx. T | Approx. Sig. |
|---|---|---|---|---|---|
| Lambda | Symmetric | .011 | .009 | 1.180 | .238 |
|  | Happy Dependent | .000 | .000 | .(b) | .(b) |
|  | Marital Dependent | .021 | .018 | 1.180 | .238 |

a Not assuming the null hypothesis.
b Cannot be computed because the asymptotic standard error equals zero.

**Symmetric Measures**

|  | Value | Approx. Sig. |
|---|---|---|
| Phi | .280 | .000 |
| Cramer's V | .198 | .000 |
| N of Valid Cases | 999 |  |

a Not assuming the null hypothesis.
b Using the asymptotic standard error assuming the null hypothesis.

4. *Association:* Looking at the chi square output, below, we see that gender and attitudes toward sex education are independent (sig = 0.201, so we fail to reject the null hypothesis of independence). The evidence suggests that men and women hold different views on sex education.

*Pattern:* Using the data below, we see that 89.5% of the men and 92.3% of women favor sex education. 10.5% of men and 7.7% of women oppose it. In general, females were slightly more likely to favor sex education than men (and men are more likely to oppose it than women).

*Strength:* Lambda equals .000 again, indicating a weak to no relationship. Phi indicates that a weak to no relationship.

**Crosstabulation**

|  |  |  | Sex | | Total |
|---|---|---|---|---|---|
|  |  |  | MALE | FEMALE | |
| Sexeduc | Favor | Count | 255 | 349 | 604 |
|  |  | Column % | 89.5% | 92.3% | 91.1% |
|  | Oppose | Count | 30 | 29 | 59 |
|  |  | Column % | 10.5% | 7.7% | 8.9% |
| Total |  | Count | 285 | 378 | 663 |

**Chi-Square Tests**

|  | Value | df | Asymp. Sig. (2-sided) |
|---|---|---|---|
| Pearson Chi-Square | 1.633 (a) | 1 | .201 |
| N of Valid Cases | 663 | | |

a  0 cells (.0%) have expected count less than 5. The minimum expected count is 25.36.

**Directional Measures**

|  |  | Value | Asymp. Std. Error(a) | Approx. T | Approx. Sig. |
|---|---|---|---|---|---|
| Lambda | Symmetric | .003 | .022 | .130 | .896 |
| | Sexeduc Dependent | .000 | .000 | .(b) | .(b) |
| | Sex Dependent | .004 | .027 | .130 | .896 |

a  Not assuming the null hypothesis.
b  Cannot be computed because the asymptotic standard error equals zero.

**Symmetric Measures**

|  | Value | Approx. Sig. |
|---|---|---|
| Phi | -.050 | .201 |
| Cramer's V | .050 | .201 |
| N of Valid Cases | 663 | |

a  Not assuming the null hypothesis.
b  Using the asymptotic standard error assuming the null hypothesis.

5.  *Association:* A chi square hypothesis test indicates that the variables are dependent (sig=0.005, so we reject the null hypothesis of independence).

*Pattern:* Of the males, 23.8% believe that sex before marriage is always wrong, compared to 29.3% of the females. Similarly, 52.1% of the males believe that sex before marriage is not wrong at all, compared to 39.9% of the females. In general, women are very slightly more likely to view premarital sex as wrong, and men are slightly more likely to view it a not wrong.

*Strength:* Lambda indicates that we don't improve our predictions of views on premarital sex when we have knowledge of gender (lambda=0.000). Cramer's V indicates a weak to moderate relationship between the two variables (V=0.141).

**Crosstabulation**

| | | | Sex | | Total |
|---|---|---|---|---|---|
| | | | MALE | FEMALE | |
| Premarsx | Always wrong | Count | 67 | 108 | 175 |
| | | Column % | 23.8% | 29.3% | 26.9% |
| | Almst always | Count | 12 | 34 | 46 |
| | wrong | Column % | 4.3% | 9.2% | 7.1% |
| | Sometimes | Count | 56 | 79 | 135 |
| | wrong | Column % | 19.9% | 21.5% | 20.8% |
| | Not wrong at | Count | 147 | 147 | 294 |
| | all | Column % | 52.1% | 39.9% | 45.2% |
| Total | | Count | 282 | 368 | 650 |

**Chi-Square Tests**

| | Value | df | Asymp. Sig. (2-sided) |
|---|---|---|---|
| Pearson Chi-Square | 12.893 (a) | 3 | .005 |
| N of Valid Cases | 650 | | |

a  0 cells (.0%) have expected count less than 5. The minimum expected count is 19.96.

**Directional Measures**

| | | Value | Asymp. Std. Error(a) | Approx. T | Approx. Sig. |
|---|---|---|---|---|---|
| Lambda | Symmetric | .000 | .027 | .000 | 1.000 |
| | Premarsx Dependent | .000 | .000 | .(b) | .(b) |
| | Sex Dependent | .000 | .061 | .000 | 1.000 |

a  Not assuming the null hypothesis.
b  Cannot be computed because the asymptotic standard error equals zero.

**Symmetric Measures**

| | Value | Approx. Sig. |
|---|---|---|
| Phi | .141 | .005 |
| Cramer's V | .141 | .005 |
| N of Valid Cases | 650 | |

a  Not assuming the null hypothesis.
b  Using the asymptotic standard error assuming the null hypothesis.

# CHAPTER THIRTEEN
## Association between Variables Measured at the Ordinal Level

*Learning Objectives:* By the end of this chapter, you will be able to:

1.  *Calculate and interpret gamma and Spearman's rho.*
2.  *Explain the logic of proportional reduction in error in terms of gamma.*
3.  *Use gamma and Spearman's rho to analyze and describe a bivariate relationship in terms of the three questions introduced in Chapter 12.*

## KEY TERMS

✓  Gamma (G)  (p. 308)
✓  $N_s$  (p. 309)
✓  $N_d$  (p. 310)
✓  Spearman's rho ($r_s$)  (p. 308)

## CHAPTER SUMMARY

In this chapter, you are introduced to measures of association for ordinal variables.  Two important, yet distinct, measures under this heading are gamma and Spearman's rho.

## CHAPTER OUTLINE

### 13.1  Introduction.

*   An important point to remember for measures of associations is that they are simply descriptions of the relationships that appear in our sample.  If we want to infer our sample results to the rest of the population, we need to conduct either a Z-test (for gamma) or a t-test (for Spearman's rho).  These tests will indicate whether our sample results can be inferred to the rest of the population (it is a significant relationship) or not.  However, hypotheses tests for gamma and Spearman's rho are not covered in your textbook.

### 13.2  Proportional Reduction in Error (PRE).

*   Previously, when analyzing nominal variables, PRE measures compared two "predictions," and the interpretation focused on how the value on one variable was predicted from the value of another variable.
*   For gamma, the predictions now focus on predicting the order of pairs of cases on one variable from the order of pairs on the other variable.
*   For Spearman's rho, the predictions focus on how the ranks of scores of the cases on one variable can be predicted from the ranks of scores on the other variable.

### 13.3 Gamma.

- Gamma (G) is used when you have two ordinal variables (or one ordinal and one interval-ratio) in a bivariate table of only a few categories (less than six categories per variable). This type of data, also called <u>collapsed ordinal data</u>, is used quite often in social science research, such as in scales which range from Strongly Agree to Strongly Disagree, or in social class categories, such as lower, working, middle, and upper.

- Gamma, which is used for ordinal data in a bivariate table, is a funny statistic to grasp, but bear with me and we'll get through it. Basically it works like this: when we used PRE measures for nominal data we were trying to predict one variable with knowledge of another variable (think back to the example of predicting a person's height with knowledge of the independent variable, gender). In gamma, you are still trying to predict the dependent variable with knowledge of the independent variable, but now you are trying to do so with *pairs* of cases.

- Think of it this way: Ruth and Cindy are both on a swim team. If we want to predict who is in the faster swim lane (the six lanes are ordered from fastest to slowest) with no other information, we have a one-in-three chance of correctly guessing (they are either in the same lane, Ruth is faster, or Cindy is faster). Let's say, though, that we know how often they each work out: Cindy works out a lot, Ruth works out fairly often. Given this information, we know that Cindy is higher on the independent variable (amount of workouts) than Ruth, therefore, we can predict that she will also be higher on the dependent variable (the swim lane). By utilizing pairs of cases (Ruth and Cindy) on both variables, we can better predict the outcome of the dependent variable.

- Gamma utilizes all pairs of cases ranked the same ($N_s$) and all pairs of cases ranked differently ($N_d$) to determine the strength of the association. Cindy works out more than Ruth and is in a faster swim lane, thus the pair of Cindy and Ruth is considered $N_s$, or ranked the same on both variables. If we compare Cindy with David, who is a natural swimmer and who works out less frequently but is still in a faster lane than Cindy, the pair would be ranked $N_d$, or differently. Gamma compares the number of cases ranked the same to the number of cases ranked differently to determine both the strength and the direction of the relationship between the two variables.

- Gamma is also a PRE measure, therefore, we can refer to reducing the amount of errors in predicting the order of pairs on the dependent variable with knowledge of the independent variable. Your textbook suggests that gamma values between 0 and 0.30 can be described as <u>weak</u>, values between .31 and .60 as <u>moderate</u>, and values greater than .60 as <u>strong</u>.

- A very important point to remember when calculating gamma is to look at the way the data are set up in the bivariate table. Be sure that the low-lows are together in the upper-left corner of the table and begin your calculations from there.

### 13.4 Determining the Direction of Relationships.

- When we have two ordinal level variables, we can ask the questions: (1) Is there an association? (2) How strong is the association? and (3) What is the pattern or direction of the association? Unlike nominal data, ordinal data are ranked and ordered, therefore we can refer to the relationship between two ordinal variables in terms of direction: Is the relationship positive or negative?

- A positive relationship, again, is one in which both variables vary in the same direction; they both increase and decrease together.

- A negative relationship, then, is one in which the variables vary in opposite directions; as one increases, the other decreases.

- Unlike nominal measures of association (phi, Cramer's V, lambda), ordinal measures of association indicate both the strength and direction of the bivariate relationship. The sign of the statistic (positive or negative) suggests the direction of the relationship (positive or negative, respectively). However, because coding schemes can be arbitrary for ordinal variables, you should always check the coding scheme before interpreting the direction of the relationship.

### 13.5 Spearman's Rho ($r_s$).

- Spearman's rho ($r_s$), on the other hand, is used when you have <u>continuous ordinal data</u>, or data that looks like interval-ratio data. For example, scales with a wide range of possibilities, such as prestige scales which range from 10-89 are not true interval-ratio variables, yet it would be difficult to break this scale into five or six categories. Thus we have a statistic that looks and acts a lot like a correlation (discussed in chapter 14) but can be used for ordinal data.

- Spearman's rho uses continuous ordinal data. This kind of data will often look like interval-ratio data, but it's not. There usually is no true zero-point, and the distance between two scores is not necessarily known. Spearman's rho gets around this by using *ranks*. To calculate rho, you must first rank the variables from highest score to lowest score. (The highest score on each variable is assigned a rank of 1.)

- Let's say we have a group of swimmers in which we know how often they work out and want to know where they placed in a recent meet. Here's how the data look:

| Swimmer | Workouts / Week | Race Finish |
|---------|-----------------|-------------|
| Carl | 7 | 1 |
| Cindy | 6 | 4 |
| Diane | 9 | 3 |
| Marlene | 6 | 5 |
| Ruth | 4 | 6 |
| Steve | 12 | 2 |
| Tom | 3 | 7 |

- We can rank how often each of the swimmers works out, as well as where they finished in the race. Note, though, that the independent variable (workouts per week) is interval-ratio. Because the other variable is ordinal, though, we must use an ordinal measure of association. Always use the measure appropriate for the lowest level of measurement between the two variables.
- We can see from these data that Steve works out the most, therefore we assign him the rank of one on the first variable. Carl came in first in the race, so he gets the rank of one on the second variable. We continue on in this manner until every person receives a rank on the first variable and a rank on the second variable. These ranks are then compared, and the rho statistic is computed.
- Spearman's rho can range from -1 to +1 indicating the strength and direction of the relationship. Like gamma, a value of -1 indicates a perfect negative relationship and a value of +1 indicates a perfect positive relationship. Unfortunately, rho by itself cannot be interpreted as a PRE measure. Instead you must square rho ($r_s^2$) for it to be interpreted as such. Rho and Gamma are symmetric measures; the calculated value will not change depending on which variable is designated as dependent.
- The rho for the above data is +0.88 and $r_s^2=0.77$. Therefore, we would conclude that there is a fairly strong, positive association between these two variables (swimmers who work out more often finish higher in the race). We reduce the amount of errors in predicting the rank of finishing place of the swimmers in the swim meet by 77% with knowledge of the rank of number of times they work out per week. (When interpreting $rho^2$, make sure to make reference to the ranks.)

## MULTIPLE-CHOICE QUESTIONS

1. Gamma is an appropriate measure for _____ level of measurement variables.
   a. nominal
   b. ordinal
   c. interval-ratio
   d. all of these choices

   *Learning Objective: 1*

2. Gamma uses _____ to determine association.
   a. means
   b. populations
   c. row marginals
   d. pairs of cases

   *Learning Objective: 1*

3. Which of the measures listed below is not based on the logic of proportional reduction in error?
   a. rho
   b. lambda
   c. rho$^2$
   d. gamma

   *Learning Objective: 2*

4. Gamma is appropriate for _____ ordinal data, whereas Spearman's rho is appropriate for _____ ordinal data.
   a. ordinal, interval-ratio
   b. collapsed, continuous
   c. bivariate, multivariate
   d. continuous, collapsed

   *Learning Objective: 1*

5. The term $N_s$ is the number of pairs of cases ranked _____ on the two variables.
   a. simultaneously
   b. systematically
   c. the same
   d. symbolically

   *Learning Objective: 3*

6. The term $N_d$ is the number of pairs of cases ranked _____ on the two variables.
   a. deviations
   b. diagonally
   c. directionally
   d. differently

   *Learning Objective: 3*

7. If the cases on one variable are predicted without error from another variable, we call this a _____ relationship.
   a. null
   b. perfect
   c. positive
   d. negative

   *Learning Objective: 3*

8.  When low scores on one variable are associated with high values on the other variable, the relationship is _____.
    a. negative
    b. asymmetric
    c. perfect
    d. positive

    *Learning Objective: 3*

9.  Gamma ranges from _____ to _____.
    a. 0, +1
    b. −1, 0
    c. −1, +1
    d. 0, infinity

    *Learning Objective: 1*

10. Spearman's rho ranges from _____ to _____.
    a. 0, +1
    b. −1, +1
    c. 0, infinity
    d. −1, 0

    *Learning Objective: 1*

11. A Spearman's rho value of +.40 indicates that we would reduce our errors of prediction by _____ percent.
    a. .40
    b. 60
    c. 16
    d. 40

    *Learning Objective: 2*

12. A gamma value of -.60 indicates that we would reduce our errors of prediction by _____ percent.
    a. 36
    b. 40
    c. 60
    d. .60

    *Learning Objective: 2*

13. Spearman's rho uses _____ to determine if two variables are associated.
    a. means
    b. standard deviations
    c. ranks on both variables
    d. collapsed data

    *Learning Objective: 1*

14. The sum of D in Spearman's rho will always equal _____.
    a. 0
    b. 1
    c. the mean
    d. Z

    *Learning Objective: 3*

15. If the two variables are measured at different levels, you should use the statistics for the _____ measure of association.
    a. higher
    b. lower
    c. ordinal
    d. tau-b

    *Learning Objective: 1*

16. A gamma of _____ indicates a perfect negative relationship.
    a. 0
    b. -1
    c. +1
    d. ±1

    *Learning Objective: 1*

17. Assume that our dependent variable has three categories (low, medium, high) and that our independent variable also has three categories (low, medium, high). Starting from the "LM" cell ("low" on the dependent variable and "medium" on the independent variable), which of the cells listed below would not form "same ranked" pairs?
a. LL
b. MM
c. MH
d. LH

*Learning Objective: 3*

18. Assume that our dependent variable has three categories (low, medium, high) and that our independent variable also has three categories (low, medium, high). Starting from the "LM" cell ("low" on the dependent variable and "medium" on the independent variable), which of the cells listed below would not form "different ranked" pairs?
a. LL
b. MM
c. MH
d. LH

*Learning Objective: 3*

19. Of the gamma values listed below, which would indicate a stronger relationship than a gamma of -.40.
a. -.50
b. +.40
c. +.25
d. -.11

*Learning Objective: 3*

20. Because gamma is a _____ measure of association, the value will not change depending on which variable is treated as the independent variable.
a. asymmetric
b. PRE
c. symmetric
d. collapsed ordinal variable

*Learning Objective: 1*

## WORK PROBLEMS

1. A researcher wanted to know if noise tolerance was associated with tolerance for children. She collected the following data. Calculate $N_s$ and $N_d$.

| Tolerance for Children | Noise Tolerance | | | |
|---|---|---|---|---|
| | Low | Med | High | Total |
| Low | 31 | 5 | 6 | 42 |
| Med | 14 | 23 | 15 | 52 |
| High | 7 | 15 | 17 | 39 |
| Total | 52 | 43 | 38 | 133 |

*Learning Objective: 3*

2. Using the data from the previous problem, determine the strength of the association.

*Learning Objective: 3*

3. Is intelligence related to how well a person will do in statistics class?

| Success in Stats Class | Intelligence | | | |
|---|---|---|---|---|
| | Low | Med | High | Total |
| Low | 10 | 7 | 8 | 25 |
| Med | 11 | 12 | 13 | 36 |
| High | 9 | 11 | 9 | 29 |
| Total | 30 | 30 | 30 | 90 |

*Learning Objective: 3*

4. Media campaigns were used in several developing countries to promote family planning. Researchers were brought in to find out if these campaigns were a success. Calculate $N_s$ and $N_d$.

| Knowledge of Family Planning | Media Consumption | | |
|---|---|---|---|
| | Low | High | Total |
| Low | 262 | 71 | 333 |
| Med | 74 | 197 | 271 |
| High | 43 | 275 | 318 |
| Total | 379 | 543 | 922 |

*Learning Objective: 3*

5. Using the data from the previous problem, determine if media consumption associated with knowledge of family planning methods?

*Learning Objective: 3*

6. Is the popularity of an older sibling associated with the popularity of the younger sibling? Sixty-two younger siblings were asked to rate the popularity of their older sibling and their own popularity. How strong is the association?

| Popularity of Younger Sibling | Popularity of Older Sibling | | |
| --- | --- | --- | --- |
| | Low | High | Total |
| Low | 21 | 10 | 31 |
| High | 14 | 17 | 31 |
| Total | 35 | 27 | 62 |

*Learning Objective: 3*

7. Calculate D and $D^2$ for the following data.

| Respondent | Class Ranking | Popularity Ranking |
| --- | --- | --- |
| 1 | 1 | 10 |
| 2 | 7 | 5 |
| 3 | 2 | 2 |
| 4 | 5 | 6 |
| 5 | 11 | 7 |
| 6 | 4 | 12 |
| 7 | 10 | 9 |
| 8 | 3 | 1 |
| 9 | 6 | 4 |
| 10 | 8 | 8 |
| 11 | 12 | 11 |
| 12 | 9 | 3 |

*Learning Objective: 3*

8. Using data from the previous problem, determine if class ranking associated with popularity ranking?

*Learning Objective: 3*

9.  Is income associated with attitudes toward welfare?

| Attitude toward Welfare | Income | | | |
|---|---|---|---|---|
| | Low | Moderate | High | Total |
| Strongly Oppose | 58 | 57 | 68 | 183 |
| Oppose | 69 | 71 | 56 | 196 |
| No Opinion | 60 | 121 | 26 | 207 |
| Favor | 75 | 149 | 40 | 264 |
| Strongly Favor | 63 | 52 | 35 | 150 |
| Total | 325 | 450 | 225 | 1000 |

*Learning Objective: 3*

10. Much research has been conducted in the field of stratification. Specifically, the question has been asked: Does father's occupational prestige affect son's occupational prestige? Using the data below, how strongly associated are these two variables?

| Respondent | Father's Prestige Score | Son's Prestige Score |
|---|---|---|
| 1 | 50 | 62 |
| 2 | 25 | 31 |
| 3 | 87 | 56 |
| 4 | 33 | 47 |
| 5 | 46 | 51 |
| 6 | 86 | 86 |
| 7 | 52 | 57 |
| 8 | 48 | 50 |
| 9 | 71 | 67 |
| 10 | 86 | 89 |
| 11 | 75 | 32 |
| 12 | 23 | 74 |
| 13 | 30 | 34 |
| 14 | 18 | 46 |
| 15 | 80 | 72 |

*Learning Objective: 3*

11. Is the amount of effort you put into your schoolwork associated with grades?

| Grades | Effort in Schoolwork | | |
|---|---|---|---|
| | Low | High | Total |
| Low | 37 | 32 | 69 |
| High | 7 | 14 | 21 |
| Total | 44 | 46 | 90 |

*Learning Objective: 3*

12. It has been said that people rarely marry above "their station." Below are some data regarding the SES of the respondent's family (their "station") and the SES of their spouse's family. Is this statement true?

| SES of Spouse's Family | SES of Family | | | | |
|---|---|---|---|---|---|
| | Upper | Middle | Working | Lower | Total |
| Upper | 221 | 71 | 21 | 4 | 317 |
| Middle | 20 | 560 | 370 | 33 | 983 |
| Working | 7 | 203 | 420 | 100 | 730 |
| Lower | 2 | 166 | 189 | 113 | 470 |
| Total | 250 | 1000 | 1000 | 250 | 2500 |

*Learning Objective: 3*

13. The number of friends a person has is hypothesized to have a positive effect on his/her quality of life. Given the data below, how strong is the relationship between these two variables?

| Respondent | Number of Friends | Quality of Life Score |
|---|---|---|
| John | 5 | 7 |
| Mary | 9 | 10 |
| Jane | 7 | 6 |
| Susan | 3 | 5 |
| George | 1 | 3 |
| Jeff | 10 | 10 |
| Claire | 4 | 5 |
| Stuart | 15 | 8 |
| Sheryl | 6 | 9 |
| Emily | 3 | 6 |
| Bill | 5 | 7 |

*Learning Objective: 3*

## SPSS or MicroCase WORK PROBLEMS

The following work problems can be completed using either SPSS or MicroCase. The solutions will be the same regardless of which one you use. Your instructor may have a preference, so be sure to check with him/her prior to beginning these problems.

1.   Using the GSS 2006 dataset, is there a relationship between the variables how often a person attends religious services (*attend*) and his/her education (*degree*)? How strong is this relationship and in what direction?

2.   It is hypothesized that people with higher prestige (*sei*) watch fewer hours of television a day (*tvhours*) than people with lower prestige. Discuss the strength and direction of this relationship.

3.   What is the relationship between education (*degree*) and income (*income3*). Does this relationship exist, how strong is it, and in what direction?

4.   What is the relationship between education (*degree*) and attitudes toward sex education (*sexeduc*). Does this relationship exist, how strong is it, and in what direction?

5.   What is the relationship between education (*degree*) and confidence in the educational system (*coneduc*). Does this relationship exist, how strong is it, and in what direction?

## ANSWER KEY

## MULTIPLE-CHOICE QUESTIONS

| | | | | | | |
|---|---|---|---|---|---|---|
| 1. | b | (p. 308) | 11. | c | (p. 320) |
| 2. | d | (p. 309-310) | 12. | c | (p. 309) |
| 3. | a | (p. 308, 320) | 13. | c | (p. 317-318) |
| 4. | b | (p. 308) | 14. | a | (p. 319) |
| 5. | c | (p. 309) | 15. | b | (p. 308) |
| 6. | d | (p. 310) | 16. | b | (p. 309) |
| 7. | b | (p. 309) | 17. | c | (p. 309-310) |
| 8. | a | (p. 313-315) | 18. | d | (p. 311-312) |
| 9. | c | (p. 309) | 19. | a | (p. 309) |
| 10. | b | (p. 319) | 20. | c | (p. 313) |

## WORK PROBLEMS

1.

$$N_s= \quad 31(23+15+15+17) \quad = 2170$$
$$+ 14(15+17) \quad + \quad 448$$
$$+ 5(15+17) \quad + \quad 160$$
$$+ 23(17) \quad + \quad 391$$
$$\overline{\quad\quad 3169}$$

$$N_d= \quad 6(23+15+14+7) \quad = \quad 354$$
$$+ 15(15+7) \quad + \quad 330$$
$$+ 5(14+7) \quad + \quad 105$$
$$+ 23(7) \quad + \quad 161$$
$$\overline{\quad\quad 950}$$

2. We use gamma because we have collapsed ordinal data.

$$G = \frac{N_s - N_d}{N_s + N_d} = \frac{3169 - 950}{3160 + 950} = 0.54$$

Therefore, we reduce our errors by 54% when predicting tolerance of children by first knowing tolerance of noise. This is a moderate positive relationship. As one's noise tolerance increases, so does tolerance for children.

3. We use gamma because we have collapsed ordinal data.

$$N_s= \quad 10(12+11+13+9) \quad = \quad 450$$
$$+ 11(11+9) \quad + \quad 220$$
$$+ 7(13+9) \quad + \quad 154$$
$$+ 12(9) \quad + \quad 108$$
$$\overline{\quad\quad 932}$$

$$N_d= \quad 8(12+11+11+9) \quad = \quad 344$$
$$+ 13(11+9) \quad + \quad 260$$
$$+ 7(11+9) \quad + \quad 140$$
$$+ 12(9) \quad + \quad 108$$
$$\overline{\quad\quad 852}$$

$$G = \frac{N_s - N_d}{N_s + N_d} = \frac{932 - 852}{932 + 852} = 0.04$$

Our gamma is very weak; therefore, intelligence has very little to do with how well one will do in statistics course. In fact, only four percent of errors are reduced in predicting class success with the variable intelligence. Even though the relationship is quite weak, gamma suggests a positive relationship; more intelligent students perform better in their statistics courses.

4.

$$N_s= \quad 262(197+275) \qquad = \quad 123664$$
$$+ \ 74(275) \qquad\qquad + \quad \underline{20350}$$
$$\qquad\qquad\qquad\qquad\qquad\qquad 144014$$

$$N_d= \quad 71(74+43) \qquad = \qquad 8307$$
$$+ \ 197(\ 43) \qquad\qquad + \quad \underline{8471}$$
$$\qquad\qquad\qquad\qquad\qquad\qquad 16778$$

5.  We use gamma because we have collapsed ordinal data.

$$G = \frac{N_s - N_d}{N_s + N_d} = \frac{144014 - 16778}{144014 + 16778} = 0.79$$

Yes, media consumption is associated with knowledge of family planning. We can reduce our error in predicting family planning knowledge by first knowing the amount of media consumed by 79%. This gamma indicates a strong, positive bivariate relationship. As media consumption increases, family planning knowledge also increases.

6.  We use gamma because we have collapsed ordinal data.

$$N_s = 21(17) = \ 357$$

$$N_d = 10(14) = \ 140$$

$$G = \frac{N_s - N_d}{N_s + N_d} = \frac{357 - 140}{357 + 140} = 0.44$$

There seems to be a moderate, positive association between these two variables. We reduce our errors of prediction by 44% with knowledge of the popularity of the older sibling. Younger siblings with popular older siblings are also more popular.

7.

| Respondent | Class Ranking | Popularity Ranking | D | D² |
|---|---|---|---|---|
| 1 | 1 | 10 | (1-10=) -9 | (9²=) 81 |
| 2 | 7 | 5 | (7-5=)  2 | (2²=)  4 |
| 3 | 2 | 2 | 0 | 0 |
| 4 | 5 | 6 | -1 | 1 |
| 5 | 11 | 7 | 4 | 16 |
| 6 | 4 | 12 | -8 | 64 |
| 7 | 10 | 9 | 1 | 1 |
| 8 | 3 | 1 | 2 | 4 |
| 9 | 6 | 4 | 2 | 4 |
| 10 | 8 | 8 | 0 | 0 |
| 11 | 12 | 11 | 1 | 1 |
| 12 | 9 | 3 | 6 | 36 |
|  |  |  | ΣD = 0 | ΣD² = 212 |

8.    We use Spearman's rho because we have continuous ordinal data.

$$r_s = 1 - \frac{6 \sum D^2}{N(N^2 - 1)} = 1 - \frac{6(212)}{12(143)} = 1 - 0.76 = 0.26$$

Therefore, there is a weak, but positive association between class ranking and popularity. We reduce our errors by only 7% ($r_s^2 = 0.07$) in predicting popularity ranking by first knowing class ranking. Respondents with higher class rankings also have higher popularity rankings.

9.     We use gamma because we have collapsed ordinal data.

$$
\begin{aligned}
N_s= \quad & 58(71+121+149+56+52+26+40+35) & = \quad & 31900 \\
& + 69(121+149+52+26+40+35) & + \quad & 29187 \\
& + 60(149+52+26+40+35) & + \quad & 18120 \\
& + 75(52+35) & + \quad & 6525 \\
& + 57(56+26+40+35) & + \quad & 8949 \\
& + 71(26+40+35) & + \quad & 7171 \\
& + 121(40+35) & + \quad & 9075 \\
& + 149(35) & + \quad & \underline{5215} \\
& & & 116142
\end{aligned}
$$

$$
\begin{aligned}
N_d= \quad & 68(71+121+149+52+69+60+75+63) & = \quad & 44880 \\
& + 56(121+149+52+60+75+63) & + \quad & 29120 \\
& + 26(149+52+75+63) & + \quad & 8814 \\
& + 40(52+63) & + \quad & 4600 \\
& + 57(69+60+75+63) & + \quad & 15219 \\
& + 71(60+75+63) & + \quad & 14058 \\
& + 121(75+63) & + \quad & 16698 \\
& + 149(63) & + \quad & \underline{9387} \\
& & & 142776
\end{aligned}
$$

$$
G = \frac{N_s - N_d}{N_s + N_d} = \frac{116142 - 142776}{116142 + 142776} = -0.103
$$

There is a weak, negative association between income and attitudes toward welfare. As income increases, attitudes toward welfare become less favorable. We reduce our error in predicting attitude about welfare by 10% with knowledge of income.

10. We use Spearman's rho because we have continuous ordinal data.

| Father's Prestige Score | Rank | Son's Prestige Score | Rank | D | D² |
|---|---|---|---|---|---|
| 50 | 8 | 62 | 6 | 2 | 4 |
| 25 | 13 | 31 | 15 | -2 | 4 |
| 87 | 1 | 56 | 8 | -7 | 49 |
| 33 | 11 | 47 | 11 | 0 | 0 |
| 46 | 10 | 51 | 9 | 1 | 1 |
| 86 | 2.5 | 86 | 2 | 0.5 | 0.25 |
| 52 | 7 | 57 | 7 | 0 | 0 |
| 48 | 9 | 50 | 10 | -1 | 1 |
| 71 | 6 | 67 | 5 | 1 | 1 |
| 86 | 2.5 | 89 | 1 | 1.5 | 2.25 |
| 75 | 5 | 32 | 14 | -9 | 81 |
| 23 | 14 | 74 | 3 | 11 | 121 |
| 30 | 12 | 34 | 13 | -1 | 1 |
| 18 | 15 | 46 | 12 | 3 | 9 |
| 80 | 4 | 72 | 4 | 0 | 0 |
| | | | | $\Sigma D = 0$ | $\Sigma D^2 = 274.5$ |

$$r_s = 1 - \frac{6 \sum D^2}{N(N^2 - 1)} = 1 - \frac{6(274.5)}{15(224)} = 1 - 0.49 = 0.51$$

There is a moderate positive relationship between these two variables; the higher father's occupational prestige, the higher son's occupational prestige. We reduce our errors by 26% ($r_s^2 = 0.26$) when predicting son's prestige ranking given father's prestige ranking.


11. We use gamma because we have collapsed ordinal data.

$N_s = 37(14) = 518$

$N_d = 32(7) = 224$

$$G = \frac{N_s - N_d}{N_s + N_d} = \frac{518 - 224}{518 + 224} = 0.40$$

There is a moderate positive association between these two variables; greater schoolwork effort produces higher grades. We reduce our errors by 40% when predicting grades with schoolwork effort.

12. We use gamma because we have collapsed ordinal data.

$$
\begin{aligned}
N_s = \quad & 221(560+203+166+370+420+189+33+100+113) & = & \quad 476034 \\
& + 20(203+166+420+189+100+113) & + & \quad 23820 \\
& + 7(166+189+113) & + & \quad 3276 \\
& + 71(370+420+189+33+100+113) & + & \quad 86975 \\
& + 560(420+189+100+113) & + & \quad 460320 \\
& + 203(189+113) & + & \quad 61306 \\
& + 21(33+100+113) & + & \quad 5166 \\
& + 370(100+113) & + & \quad 78810 \\
& + 420(113) & + & \quad 47460 \\
\hline
& & & \quad 1243167
\end{aligned}
$$

$$
\begin{aligned}
N_d = \quad & 4(370+420+189+560+203+166+20+7+2) & = & \quad 7748 \\
& + 33(420+89+203+166+7+2) & + & \quad 29040 \\
& + 100(189+166+2) & + & \quad 35700 \\
& + 21(560+203+166+20+7+2) & + & \quad 20118 \\
& + 370(203+166+7+2) & + & \quad 139860 \\
& + 420(166+2) & + & \quad 70560 \\
& + 71(20+7+2) & + & \quad 2059 \\
& + 560(7+2) & + & \quad 5040 \\
& + 203(2) & + & \quad 406 \\
\hline
& & & \quad 310531
\end{aligned}
$$

$$
G = \frac{N_s - N_d}{N_s + N_d} = \frac{1243167 - 310531}{1243167 + 310531} = 0.60
$$

Therefore, there is a moderate (almost strong) positive relationship between these two variables; the higher one's "station," the higher they marry. We reduce our error by 60% in predicting the SES of the family a person will marry into by first knowing the SES of his/her family.

13. We use Spearman's rho because we have continuous ordinal data.

| Number of Friends | Rank | Quality of Life Score | Rank | D | D² |
|---|---|---|---|---|---|
| 5 | 6.5 | 7 | 5.5 | 1 | 1 |
| 9 | 3 | 10 | 1.5 | 1.5 | 2.25 |
| 7 | 4 | 6 | 7.5 | -3.5 | 12.25 |
| 3 | 9.5 | 5 | 9.5 | 0 | 0 |
| 1 | 11 | 3 | 11 | 0 | 0 |
| 10 | 2 | 10 | 1.5 | 0.5 | 0.25 |
| 4 | 8 | 5 | 9.5 | -1.5 | 2.25 |
| 15 | 1 | 8 | 4 | -3 | 9 |
| 6 | 5 | 9 | 3 | 2 | 4 |
| 3 | 9.5 | 6 | 7.5 | 2 | 4 |
| 5 | 6.5 | 7 | 5.5 | 1 | 1 |
|  |  |  |  | $\Sigma D = 0$ | $\Sigma D^2 = 36$ |

$$r_s = 1 - \frac{6 \sum D^2}{N(N^2 - 1)} = 1 - \frac{6(36)}{11(120)} = 1 - 0.16 = 0.84$$

Therefore there is a fairly strong, positive relationship between these two variables; one's quality of life increases with the number of friends. When we square this number, we find that we make 71% fewer errors in predicting ranks on quality of life with knowledge of ranks on number of friends.

## SPSS or MicroCase WORK PROBLEMS

If you had problems with any of these questions, be sure to re-read the appropriate sections at the end of the chapter in your textbook.

1. People with a high school diploma are slightly more likely to attend religious services less than once a year (32.2%); those with graduate degrees are more likely to attend religious services once a year to several times a year (30.3%); people with less than high school are very slightly more likely to attend religious serves once a month to almost weekly (29.0%); and people with bachelor degrees are more likely to attend religious services every week or more (30.5%). This relationship can be described as positive because income increases with education. Gamma (G=0.076) indicates a weak positive relationship between the two variables; as education increases, religious service attendance increases slightly. We reduce our errors in predicting religious service attendance by 7.6% when we have knowledge of educational category.

**Crosstabulation**

| | | RS HIGHEST DEGREE | | | | | |
|---|---|---|---|---|---|---|---|
| | | LT HIGH SCHOOL | HIGH SCHOOL | JUNIOR COLLEGE | BACHELOR | GRADUATE | Total |
| LESS THAN 1 A YEAR | Count | 68 | 247 | 27 | 65 | 34 | 441 |
| | Column % | 30.8% | 32.2% | 23.9% | 24.8% | 25.8% | 29.5% |
| 1 A YR-SEV TIMES WK | Count | 49 | 173 | 31 | 54 | 40 | 347 |
| | Column % | 22.2% | 22.6% | 27.4% | 20.6% | 30.3% | 23.2% |
| 1 A MNTH-ALMST WKLY | Count | 64 | 150 | 21 | 63 | 24 | 322 |
| | Column % | 29.0% | 19.6% | 18.6% | 24.0% | 18.2% | 21.6% |
| EVERY WEEK + | Count | 40 | 196 | 34 | 80 | 34 | 384 |
| | Column % | 18.1% | 25.6% | 30.1% | 30.5% | 25.8% | 25.7% |
| Total | | 221 | 766 | 113 | 262 | 132 | 1494 |

**Symmetric Measures**

| | Value | Asymp. Std. Error(a) | Approx. T(b) | Approx. Sig. |
|---|---|---|---|---|
| Gamma | .076 | .030 | 2.550 | .011 |
| N of Valid Cases | 1494 | | | |

a  Not assuming the null hypothesis.
b  Using the asymptotic standard error assuming the null hypothesis.

2.  Prestige (the socioeconomic index, SEI) is a continuous ordinal variable, whereas hours of television watched a day is an interval-ratio level variable. Because we have one ordinal and one interval-ratio variable, we need to use the lower measure of association, Spearman's rho. To do so, you must click on the "Analyze" menu; click on the "Correlate" option; then click on "Bivariate." Put the two variables *sei* and *tvhours* into the variables box. To conduct a Spearman's rho test, click on the box titled "Spearman" below the variables list. (Be sure only Spearman is checked).

To read this table, look for the Spearman's rho correlation between prestige and TV hours per day, which is -0.266. We reduce the number of errors in predicting someone's rank on hours of TV watched by 7.1% ($-0.266^2$) when we have knowledge of their rank on prestige. [Don't forget: Spearman's rho can be interpreted as a PRE measure only after it has been squared.] Finally, there is a negative relationship between these two variables: as prestige increases, hours of TV watched per day decreases.

**Correlations**

| | | | SEI | TVHOURS |
|---|---|---|---|---|
| Spearman's rho | SEI | Correlation Coefficient | 1.000 | -.266 |
| | | Sig. (2-tailed) | . | .000 |
| | | N | 1426 | 641 |
| | TVHOURS | Correlation Coefficient | -.266 | 1.000 |
| | | Sig. (2-tailed) | .000 | . |
| | | N | 641 | 673 |

3.    People with less than a high school diploma are the most likely to have low income (68.3%); high school and junior college graduates are more likely to have moderate income (32.7% for junior college grads and 32.0% for high school grads); and people with graduate degrees are the most likely to have high incomes (64.1%). This relationship can be described as positive because income increases with education. Gamma (G=0.532) indicates a moderate to strong relationship between the two variables. We reduce our errors in predicting income by 53.2% when we have knowledge of educational category.

**Crosstabulation**

| | | | DEGREE | | | | | Total |
|---|---|---|---|---|---|---|---|---|
| | | | LT HIGH SCHOOL | HIGH SCHOOL | JUNIOR COLLEGE | BACHELOR | GRADUATE | |
| INCOME | Low | Count | 123 | 251 | 22 | 35 | 11 | 442 |
| | | Col % | 68.3% | 38.4% | 22.4% | 15.0% | 9.4% | 34.5% |
| | Moderate | Count | 44 | 209 | 32 | 60 | 31 | 376 |
| | | Col % | 24.4% | 32.0% | 32.7% | 25.8% | 26.5% | 29.3% |
| | High | Count | 13 | 194 | 44 | 138 | 75 | 464 |
| | | Col % | 7.2% | 29.7% | 44.9% | 59.2% | 64.1% | 36.2% |
| | Total | | 180 | 654 | 98 | 233 | 117 | 1282 |

**Symmetric Measures**

| | Value | Asymp. Std. Error(a) | Approx. T(b) | Approx. Sig. |
|---|---|---|---|---|
| Gamma | .532 | .028 | 17.207 | .000 |
| N of Valid Cases | 1282 | | | |

a  Not assuming the null hypothesis.
b  Using the asymptotic standard error assuming the null hypothesis.

4. People with bachelor and graduate degrees are most likely to favor sex education in schools (95.4% and 96.5%, respectively), and people with less than a high school diploma or a junior college degree are most likely to oppose sex education (14.0% and 14.3%, respectively). The relationship between educational degree and attitudes toward sex education is negative, in that favorability to sex education increases with education. Gamma (G=-.257) indicates a weak to moderate relationship between the two variables. We reduce our errors in predicting sex education attitudes by 25.7% when we have knowledge of educational category.

**Crosstabulation**

| | | DEGREE | | | | | Total |
|---|---|---|---|---|---|---|---|
| | | LT HIGH SCHOOL | HIGH SCHOOL | JUNIOR COLLEGE | BACHELOR | GRADUATE | |
| FAVOR | Count | 74 | 303 | 48 | 124 | 55 | 604 |
| | Column % | 86.0% | 90.7% | 85.7% | 95.4% | 96.5% | 91.1% |
| OPPOSE | Count | 12 | 31 | 8 | 6 | 2 | 59 |
| | Column % | 14.0% | 9.3% | 14.3% | 4.6% | 3.5% | 8.9% |
| Total | Count | 86 | 334 | 56 | 130 | 57 | 663 |

**Symmetric Measures**

| | Value | Asymp. Std. Error(a) | Approx. T(b) | Approx. Sig. |
|---|---|---|---|---|
| Gamma | -.257 | .101 | -2.469 | .014 |
| N of Valid Cases | 663 | | | |

a  Not assuming the null hypothesis.
b  Using the asymptotic standard error assuming the null hypothesis.

5. People with less than a high school diploma are most likely to have a great deal of confidence in education (42.3%); people with a bachelors degree are most likely to have only some confidence (68.2%); and people with a graduate degree are most likely to have hardly any confidence (21.8%). The relationship is positive, in that higher educational attainment increases having hardly any confidence. Gamma (G=.102) indicates a weak relationship. We reduce our errors in predicting confidence by 10.2% when we have knowledge of educational category.

**Crosstabulation**

| | | DEGREE | | | | | Total |
|---|---|---|---|---|---|---|---|
| | | LT HIGH SCHOOL | HIGH SCHOOL | JUNIOR COLLEGE | BACHELOR | GRADUATE | |
| A GREAT DEAL | Count | 44 | 93 | 14 | 23 | 15 | 189 |
| | Col % | 42.3% | 27.8% | 24.6% | 20.9% | 27.3% | 28.6% |
| ONLY SOME | Count | 40 | 189 | 33 | 75 | 28 | 365 |
| | Col % | 38.5% | 56.4% | 57.9% | 68.2% | 50.9% | 55.2% |
| HARDLY ANY | Count | 20 | 53 | 10 | 12 | 12 | 107 |
| | Col % | 19.2% | 15.8% | 17.5% | 10.9% | 21.8% | 16.2% |
| Total | Count | 104 | 335 | 57 | 110 | 55 | 661 |

**Symmetric Measures**

| | Value | Asymp. Std. Error(a) | Approx. T(b) | Approx. Sig. |
|---|---|---|---|---|
| Gamma | .102 | .056 | 1.807 | .071 |
| N of Valid Cases | 661 | | | |

a  Not assuming the null hypothesis.
b  Using the asymptotic standard error assuming the null hypothesis.

305

# CHAPTER FOURTEEN
## Association Between Variables Measured at the Interval-Ratio Level

*Learning Objectives*: By the end of this chapter, you will be able to:

1. Interpret a scattergram.
2. Calculate and interpret slope (b), Y intercept (a), and Pearson's r and $r^2$.
3. Find and explain the least-squares regression line and use it to predict values of Y.
4. Explain the concepts of total, explained, and unexplained variance.
5. Use regression and correlation techniques to analyze and describe a bivariate relationship in terms of the three questions introduced in Chapter 12.

## KEY TERMS

- ✓ Coefficient of Determination ($r^2$)  (p. 341)
- ✓ Conditional Means of Y  (p. 335)
- ✓ Explained Variation  (p. 343)
- ✓ Linear Relationship  (p. 332)
- ✓ Pearson's r ($r$)  (p. 339)
- ✓ Regression Line  (p. 332)
- ✓ Scattergram  (p. 330)
- ✓ Slope ($b$)  (p. 336)
- ✓ Total Variation  (p. 342)
- ✓ Unexplained Variation  (p. 343)
- ✓ *Y intercept* (*a*)  (p. 336)
- ✓ Y'  (p. 334)

## CHAPTER SUMMARY

When we are interested in the association between two interval-ratio variables, we want to keep in mind the same three questions that we've seen in previous chapters:  (1) Does an association exist?  (2) How strong is the association?  and (3) What is the direction of the association?

## CHAPTER OUTLINE

**14.1  Introduction.**

## 14.2 Scattergrams.

- In order to answer the first question ("Does an association exist?") for nominal and ordinal data we used a bivariate table and compared the conditional distributions of Y. For interval-ratio variables, we use what is called a scattergram to examine the conditional distributions of Y. A scattergram is a graph that displays the values of the independent variable (X) along the horizontal axis and the values of the dependent variable (Y) along the vertical axis. The values for each respondent on both variables are then plotted on the graph. The dots above each X score are the conditional distributions of Y. The display of the plotted values gives us an idea of whether or not the variables are associated. If the conditional distributions of Y (the dots above each X value) change across different values of X, then there is an association between the two variables.

- To analyze the association between the two variables we find the best-fitting regression line. The regression line is the line that best fits the points on the scattergram. Your textbook describes a way to freehand a regression line on your scattergram. You can do this by drawing a straight line through the cluster of dots such that your line touches every dot (or comes as close as possible to touching every dot). A second way to draw a regression line is to connect the conditional means of Y, which are calculated by summing all of the Y values for each X value and then dividing by the number of such cases. However, the conditional means of Y will often not form a straight line. (In fact, they only form a straight line when the variables share a perfect relationship.) Because of this, we find the least squares regression line, or the one straight line that best fits the pattern of dots.

- We answer the second question ("How strong is the association?") by looking at the spread of the dots around the regression line. When every dot falls on the regression line, there is a perfect association between the two variables. When the dots don't fall on the regression line, but very close to the line, then there is a strong relationship between the variables. The farther away the dots fall from the regression line, the weaker the relationship.

- We can now answer the third question ("What is the direction of the association?") by looking at the angle of the regression line. If the values of the dependent variable increase with the values of the independent variable (the regression line goes from the lower-left to the upper-right), then we can say that there is a positive association between the two variables. If the values of the dependent variable decrease with the values of the independent variable (the regression line goes from the upper-left to the lower-right), then we can say that there is a negative association between the two variables. If the distribution of the plot is basically flat or there is no pattern to the plot (that is, it's a completely random scatter of points), then we conclude there is no association between these two variables.

## 14.3 Regression and Prediction.

- The equation for the least squares regression line is:

$$Y = a + bX$$

- Where a is the Y intercept, b is the slope of the line, X is a score on the independent variable, and Y is a score on the dependent variable. The Y intercept is where the regression line crosses the Y axis (that is, the value of Y when X is zero). The slope is the amount of change in Y for every one-unit change in X. X is the independent variable that you are using to predict the value on the dependent variable, Y.
- We can use the regression equation to describe the <u>strength</u> of the relationship between X and Y. The stronger the relationship between X and Y, the larger the effect X has on Y (or the larger the change in Y for a one unit change in X). Therefore, larger slopes (b) indicate stronger relationships, and smaller slopes (b) indicate weaker relationships. When the slope is zero the least squares regression line is flat, and the two variables are unrelated. Thus, whenever the slope is non-zero, the variables are associated with each other.
- The <u>direction</u> of the relationship can be gotten from the sign of the slope (b). A positive slope (b) indicates a positive relationship (as X increases, Y increases), and a negative slope (b) indicates a negative relationship (as X increases, Y decreases).
- Furthermore, we can use the regression equation to predict values of Y for any given value of X. Given a value for X, a slope (b), and a Y intercept (a), we can substitute these values in the regression equation and determine predicted Y (Y'). With regression, we are often attempting to predict the dependent variable when the independent variable is some given or known value. For example, predicting someone's income if she/he has 15 years of education.
- The regression equation, though, does not contain any PRE measures. In order to analyze the relationship in terms of PRE-style interpretations, we must use the coefficient of determination ($r^2$), which is the percent of variation in the dependent variable (Y) that can be explained by the independent variable (X). If someone has 15 years of education, is that a perfect predictor of how much money he/she will earn? No, but $r^2$ will indicate how much variation in income can be explained by education.

### 14.4 Computing *a* and *b*.

- To calculate the slope (b) use the following formula:

$$b = \frac{\sum (X - \overline{X})(Y - \overline{Y})}{\sum (X - \overline{X})^2}$$

  - The numerator of the formula is termed the covariation, where covariation indicates how much the two variables vary together.
  - The slope indicates the change in the dependent variable for every one unit change in the independent variable.
- To calculate the intercept (a) use the following formula:

$$a = \overline{Y} - b\overline{X}$$

- The intercept indicates where the regression line crosses the Y axis, or in other words, what the dependent variable equals when the independent variable equals zero.

## 14.5 The Correlation Coefficient (Pearson's *r*).
- Pearson's r, or the correlation coefficient, is a measure of the strength of the association between two interval-ratio level variables. It ranges from -1 to +1, where 1 is a perfect association (either positive or negative) and zero is no association. Bivariate relationships with Pearson's r values between 0 and .3 can be characterized as <u>weak,</u> with values between .31 and .60 as <u>moderate</u>, and with values greater than .6 as <u>strong</u>.
- To calculate the correlation coefficient (r) use the following formula:

$$ r = \frac{\sum (X - \overline{X})(Y - \overline{Y})}{\sqrt{\left[ \sum (X - \overline{X})^2 \right]\left[ \sum (Y - \overline{Y})^2 \right]}} $$

## 14.6 Interpreting the Correlation Coefficient: $r^2$.
- Pearson's r is a widely used measure, but it cannot be interpreted as a PRE measure. Instead, squaring the value makes it possible to interpret r in this manner. When interpreting $r^2$, we refer to the percent of variation explained in the dependent variable by the independent variable. Thus, if we have a correlation (r) of -0.50, we say that we have a moderately strong negative association and that 25% ($0.50^2$) of the variation in the dependent variable is explained by the independent variable. Although we can describe the strength of the relationship with the value of b (as described above), the best way to assess the strength of a relationship between two interval-ratio variables is to use Pearson's r and $r^2$.

## 14.7 The Correlation Matrix.
- Correlations (Pearson's r) are often presented in a correlation matrix, which is a table that contains the correlations for every possible pair of variables. The correlation matrix has each variable arrayed across the top and down the side; the correlation between each variable appears in the corresponding cell. Along the diagonal (in cell $X_1,X_1$ for example) is the correlation of each variable with itself, which, logically is 1.0. Many times these matrices will be only half complete because the matrix is symmetric (the correlation for $X_1$ and $X_2$ is the same as the correlation for $X_2$ and $X_1$). A three variable correlation matrix may look something like this:

|       | $X_1$ | $X_2$ | Y    |
| ----- | ----- | ----- | ---- |
| $X_1$ | 1.0   | -0.31 | 0.56 |
| $X_2$ |       | 1.0   | 0.76 |
| Y     |       |       | 1.0  |

- The correlation between $X_1$ and $X_2$ is -0.31, between $X_1$ and Y is 0.56, and between $X_2$ and Y is 0.76. Note that the lower triangle of the matrix is incomplete. You can complete it if you'd like with the same numbers that are in the upper triangle, but it would be redundant, so it's not necessary.
- The primary reason these matrices are used is for ease of reading the statistics. You need to look at this table only briefly to see the relationship between the three variables. Correlation matrices are not limited to any number of variables, but can get cumbersome when too many variables are included.

**14.8 Correlation, Regression, Level of Measurement and Dummy Variables.**
- Correlation and regression are such useful and powerful statistical techniques that they are used with data that are not strictly appropriate. This is not that much of a problem when "continuous" ordinal variables are used instead of true interval-ratio level variables. But often other kinds of variables – for example, "collapsed" ordinal or nominal variables – are of interest. In such cases, researchers can recode these variables to two categories (the first category needs to be coded as zero and the second category needs to be coded as one), creating a dummy variable. The regression coefficient for a dummy variable indicates how the "1" category differs from the "0" category on the dependent variable. Suppose that sex is the independent variable, coded "0" for men and "1" for women, and that income is the dependent variable. Further suppose that the regression coefficient for sex is -2500. This coefficient would then indicate that women (the "1" category") make $2500 less than men (the "0" category).

**MULTIPLE-CHOICE QUESTIONS**

1. In order to determine if an association exists between two interval-ratio variables, a _____ should be constructed.
   a. bivariate table
   b. distribution table
   c. scattergram
   d. regression equation

   *Learning Objective: 1*

2. To assess the strength of the relationship, a _____ should be used.
   a. scattergram
   b. bivariate table
   c. gamma
   d. Pearson's r

   *Learning Objective: 2*

3.    The PRE measure for interval-ratio variables is _____.
      a.  r
      b.  r$^2$
      c.  b
      d.  Y

*Learning Objective: 2*


4.    The term "conditional means of Y" refers to _____.
      a.  the mean of all scores on Y for each value of X
      b.  the mean of all scores on X for each value of Y
      c.  the mean of each cell frequency across the Y categories.
      d.  the mean of each cell frequency across each value of X

*Learning Objective: 2*


5.    A linear relationship is one in which the relationship between two variables approximates
      _____.
      a.  a downward trend
      b.  an arc
      c.  a positive association
      d.  a straight line

*Learning Objective: 3*


6.    In the regression equation (Y=a+bX), the term "a" is used to denote the _____.
      a.  slope
      b.  total variation explained
      c.  Y-intercept
      d.  regression line

*Learning Objective: 2*


7.    In the regression equation (Y=a+bX), the term "b" is used to denote the _____.
      a.  slope
      b.  total variation explained
      c.  Y-intercept
      d.  regression line

*Learning Objective: 2*

8. In the regression equation ($Y=a+bX$), the term "X" refers to _____.
   a. the amount of variance unexplained
   b. the total variation explained
   c. a score on the independent variable
   d. a predicted score on the dependent variable

   *Learning Objective: 2*


9. In a _____ association, all of the dots in a scattergram fall on the regression line.
   a. nonlinear
   b. perfect
   c. negative
   d. positive

   *Learning Objective: 1*


10. Unexplained variation refers to _____.
    a. the proportion of variation in Y not accounted for by X
    b. the proportion of variation in X not accounted for by Y
    c. the amount of variation explained by both the independent and dependent variables
    d. the spread of the Y scores around the mean

    *Learning Objective: 4*


11. Correlation can be used for what type of data?
    a. Two interval-ratio variables
    b. One interval-ratio and one ordinal variable
    c. One interval-ratio and one nominal variable
    d. It doesn't matter what level of measurement the variables are

    *Learning Objective: 2*


12. A _____ relationship is indicated by a regression line that is _____.
    a. negative, angled from bottom-left to top-right
    b. zero, vertical
    c. zero, horizontal
    d. positive, angled from top-left to bottom-right

    *Learning Objective: 1*

13. The _____ is the point where the regression line crosses the vertical axis.
    a. X intercept
    b. slope
    c. origin
    d. Y intercept

    *Learning Objective: 2*


14. The term Y' is used to denote _____.
    a. the predicted score on X
    b. the predicted score on Y
    c. the amount of variation explained in Y
    d. the Y-intercept

    *Learning Objective: 2*


15. When describing the PRE measure, we talk about the amount of _____ explained by the independent variable.
    a. linearity
    b. variation
    c. prediction
    d. homoscedasticity

    *Learning Objective: 4*


16. In a regression equation predicting income (measured in $1000s) with education (measured in years), the intercept (a) equals 1.5 and the regression coefficient (b) equals 4. Which of the below is the correct interpretation of the intercept (a)?
    a. for every additional year of education, income increases $4000
    b. when income equals zero, education equals 1.5
    c. for every four years of education, income increases $1000
    d. at zero years of education, income equals $1500

    *Learning Objective: 3*

17.     In a regression equation predicting income (measured in $1000s) with education
        (measured in years), the intercept (a) equals 1.5 and the regression coefficient (b) equals
        4. Which of the below is the correct interpretation of the regression coefficient (b)?
        a. for every additional year of education, income increases $4000
        b. for every additional year of education, income increases $4
        c. for every four years of education, income increases $1000
        d. at zero years of education, income equals $1500

        *Learning Objective: 3*

18.     In a regression equation predicting income (measured in $1000s) with education
        (measured in years), the coefficient of determination equals .50. Which of the below is
        the correct interpretation of the coefficient of determination?
        a. income explains 50% of the variation in education
        b. education explains 50% of the variation in income
        c. income explains 25% of the variation in education
        d. education explains 25% of the variation in income

        *Learning Objective: 3*

19.     The numerator of the formula for Pearson's r and b is called the _____.
        a. correlation
        b. slope
        c. intercept
        d. covariation

        *Learning Objective: 2*

20.     In a regression equation predicting income (measured in $1000s) with sex (measured as a
        dummy variable with 0=female and 1=male), the regression coefficient (b) equals 2
        Which of the below is the correct interpretation of the regression coefficient (b)?
        a. men have two more years of education than women
        b. men have two fewer years of education than women
        c. men have twice the education of women
        d. men have half the education of women

        *Learning Objective: 3*

**WORK PROBLEMS**

1.  After an exam, a professor asked each student to disclose the number of hours she/he studied for the exam. Each student's hours of study and grade are presented below. What is the correlation?

| Hours of Study | Exam Grade |
| --- | --- |
| 5 | 64 |
| 1 | 52 |
| 6 | 76 |
| 3 | 71 |
| 4 | 74 |
| 9 | 81 |
| 11 | 80 |
| 14 | 83 |
| 2 | 69 |
| 1 | 56 |
| 7 | 79 |
| 4 | 93 |
| 6 | 91 |
| 3 | 85 |
| 8 | 88 |
| 12 | 96 |

*Learning Objective: 2*

2.  Using the data from the previous problem, determine the regression equation.

*Learning Objective: 2*

3.  Using the data from the previous problem, determine the predicted exam grade for a student who studied 4 hours, 6 hours, 8 hours, or 10 hours.

*Learning Objective: 3*

4   Contemplating graduate school, Matt decided to conduct a survey on income and years of education. Specifically he was interested in the amount of increase in income for every additional year of education. Should he go to grad school? (Hint: For ease of calculations, income has been divided by 1000, so the value of 24 is really $24,0000, etc.)

| Education | Income (in $1000s) |
|-----------|--------------------|
| 12 | 24 |
| 10 | 4.5 |
| 16 | 41.7 |
| 16 | 32.5 |
| 18 | 47 |
| 21 | 157 |
| 22 | 32.5 |
| 14 | 28 |
| 16 | 34 |
| 18 | 50 |
| 8 | 12 |
| 23 | 11 |
| 14 | 29 |
| 15 | 27 |
| 12 | 25 |
| 13 | 24 |
| 11 | 14 |
| 12 | 17.4 |
| 16 | 87 |
| 19 | 98 |

*Learning Objective: 2*

5.   Using the data from the previous problem, determine the predicted income for someone with a high school degree (12 years of education), a college degree (16 years of education), a masters degree (18 years of education), or a doctorate degree (22 years of education).

*Learning Objective: 3*

6.    A researcher is interested in the effects of income on the number of children a couple decides to have. Is there a linear association between the two variables? (Hint: For ease of calculations, income has been divided by 1000.)

| Income (in $1000s) | Number of Children |
|---|---|
| 18 | 4 |
| 55 | 3 |
| 47 | 2 |
| 35 | 3 |
| 12 | 5 |
| 28 | 2 |
| 108 | 3 |
| 36 | 3 |
| 42 | 2 |
| 155 | 1 |
| 31 | 3 |
| 45 | 2 |
| 10 | 4 |
| 129 | 1 |
| 22 | 4 |
| 9 | 8 |

*Learning Objective: 2*

7.    A sociologist wondered if education led to the postponement of marriage. From the data below, what can you conclude?

| Education | Age Married |
|---|---|
| 10 | 16 |
| 12 | 18 |
| 15 | 21 |
| 16 | 20 |
| 12 | 22 |
| 16 | 29 |
| 18 | 24 |
| 14 | 22 |
| 12 | 25 |
| 18 | 21 |
| 16 | 24 |
| 12 | 19 |
| 11 | 17 |
| 14 | 20 |

*Learning Objective: 2*

8. Is there a linear association between number of hours worked per week and income? Income is measured in $1000s.

| Hours Work per Week | Income (in $1000s) |
| --- | --- |
| 40 | 47 |
| 75 | 157 |
| 42 | 32.5 |
| 34 | 28 |
| 40 | 34 |
| 42 | 50 |
| 26 | 12 |
| 40 | 11 |
| 34 | 29 |
| 45 | 27 |
| 42 | 25 |
| 32 | 24 |
| 37 | 14.5 |
| 46 | 41.7 |
| 43 | 32.5 |
| 45 | 47 |
| 80 | 159 |
| 39 | 32 |
| 40 | 28 |

*Learning Objective: 2*

9.    In a study of pro-social messages recalled from the television, it was hypothesized that the more messages seen (measured with hours of TV viewing per day), the more messages the respondent would recall. Do these data support this hypothesis?

| Hours of TV a Day | Message Recall |
|---|---|
| 7 | 3 |
| 6 | 2 |
| 3 | 2 |
| 2 | 0 |
| 6 | 5 |
| 5 | 4 |
| 4 | 3 |
| 6 | 2 |
| 7 | 3 |
| 3 | 2 |
| 4 | 4 |
| 5 | 1 |
| 2 | 2 |
| 6 | 5 |
| 7 | 3 |
| 9 | 4 |
| 4 | 2 |
| 4 | 1 |
| 7 | 5 |

*Learning Objective: 2*

10. Does religiosity decrease with income?  Are people with higher incomes likely to go to fewer services?  Income is measured in $1000s.

| Income (in $1000s) | Number of Religious Services a Year |
| --- | --- |
| 44 | 12 |
| 35 | 18 |
| 32 | 24 |
| 29 | 24 |
| 34 | 12 |
| 50 | 6 |
| 63 | 1 |
| 11 | 52 |
| 29 | 6 |
| 27 | 36 |
| 25 | 2 |
| 24 | 1 |
| 32 | 0 |
| 41 | 4 |
| 32 | 12 |
| 47 | 2 |
| 159 | 1 |
| 32 | 12 |
| 18 | 26 |

*Learning Objective: 2*

11. A professor noticed he had many married students in his class, some with children. Being a father himself, he wondered what effect the children would have on study habits of his married students. What do the data tell us about it?

| Number of Children | Hours of Study a Week |
| --- | --- |
| 4 | 16 |
| 0 | 10 |
| 2 | 12 |
| 2 | 11 |
| 1 | 14 |
| 6 | 20 |
| 0 | 10 |
| 1 | 12 |
| 2 | 15 |
| 4 | 16 |
| 3 | 14 |
| 2 | 17 |
| 0 | 9 |
| 0 | 6 |
| 1 | 10 |

*Learning Objective: 2*

12. Do people who live near the city take advantage of the cultural events more than people who live farther away?

| Proximity to City | Number of Cultural Events a Year |
| --- | --- |
| 5 | 25 |
| 12 | 18 |
| 25 | 12 |
| 40 | 2 |
| 17 | 3 |
| 10 | 24 |
| 19 | 6 |
| 28 | 2 |
| 36 | 0 |
| 6 | 3 |
| 17 | 4 |
| 28 | 2 |
| 19 | 1 |
| 10 | 7 |
| 27 | 1 |

*Learning Objective: 2*

13.  It seemed to a researcher that as children grow older, there were fewer and fewer family activities. Do our data support this hypothesis?

| Age of Child | Number of Family Activities a Year |
|:---:|:---:|
| 4 | 4 |
| 11 | 2 |
| 15 | 0 |
| 3 | 12 |
| 10 | 1 |
| 13 | 0 |
| 8 | 36 |
| 6 | 18 |
| 5 | 27 |
| 17 | 0 |
| 16 | 1 |
| 14 | 4 |
| 10 | 23 |
| 7 | 20 |

*Learning Objective: 2*

14. It has been said that if you want to get a lot accomplished, stay very busy. Do people who are very busy get more done per day than people who are not?

| Hours of Free Time per Day | Number of Activities per Week |
| --- | --- |
| 5 | 10 |
| 12 | 2 |
| 25 | 0 |
| 40 | 0 |
| 17 | 2 |
| 11 | 5 |
| 19 | 4 |
| 28 | 2 |
| 36 | 3 |
| 6 | 5 |
| 17 | 2 |
| 28 | 2 |
| 19 | 0 |
| 10 | 7 |
| 27 | 2 |
| 11 | 5 |
| 15 | 3 |
| 3 | 12 |
| 10 | 16 |
| 13 | 7 |
| 8 | 8 |
| 6 | 9 |
| 5 | 10 |

*Learning Objective: 2*

15. Is television linearly associated with violent acts in children? The hours of TV watching per day and the number of violent acts per day for 16 children are below.

| Number of TV Hours per Day | Number of Violent Acts |
|:---:|:---:|
| 2 | 0 |
| 3 | 1 |
| 4 | 3 |
| 2 | 0 |
| 6 | 4 |
| 7 | 3 |
| 5 | 7 |
| 4 | 2 |
| 5 | 4 |
| 1 | 1 |
| 3 | 0 |
| 7 | 0 |
| 8 | 4 |
| 2 | 2 |
| 4 | 5 |
| 7 | 4 |

*Learning Objective: 2*

## SPSS or MicroCase WORK PROBLEMS

The following work problems can be completed using either SPSS or MicroCase. The solutions will be the same regardless of which one you use. Your instructor may have a preference, so be sure to check with him/her prior to beginning these problems.

1. What is the relationship between number of siblings (*sibs*) and the number of children one has (*childs*)? How many children would you expect someone to have who has 2 siblings? What about someone who has 10 siblings?

2. Test the hypothesis that as income increases (*realinc*), the number of children one has decreases (*childs*).

3. What is the relationship between age (*age*) and number of children one has (*childs*)?

4. What is the relationship between age (*age*) and income (*realinc*)?

5. What is the relationship between age (*age*) and television viewership (*tvhours*)?

## ANSWER KEY

### MULTIPLE-CHOICE QUESTIONS

| | | | | | | |
|---|---|---|---|---|---|---|
| 1. | c | (p. 330) | 11. | a | (p. 339) |
| 2. | d | (p. 339-340) | 12. | c | (p. 333) |
| 3. | b | (p. 341) | 13. | d | (p. 336) |
| 4. | a | (p. 335) | 14. | b | (p. 342) |
| 5. | d | (p. 332-333) | 15. | b | (p. 342-342) |
| 6. | c | (p. 336) | 16. | d | (p. 338) |
| 7. | a | (p. 336) | 17. | a | (p. 337-338) |
| 8. | c | (p. 336) | 18. | b | (p. 345) |
| 9. | b | (p. 333) | 19. | d | (p. 340) |
| 10. | a | (p. 343) | 20. | a | (p. 349-350) |

### WORK PROBLEMS

1. The necessary sums and quantities for calculating the slope (b), Y intercept (a), and Pearson's r are listed below.

$$\Sigma X = 96 \qquad \Sigma Y = 1238 \qquad \Sigma XY = 7883$$
$$\Sigma X^2 = 808 \qquad \Sigma Y^2 = 98196 \qquad N = 16$$
$$\overline{X} = 6.0 \qquad \overline{Y} = 77.375$$

$$r = \frac{N\,\Sigma XY - (\Sigma X)(\Sigma Y)}{\sqrt{[N\,\Sigma X^2 - (\Sigma X)^2][N\,\Sigma Y^2 - (\Sigma Y)^2]}}$$

$$= \frac{(16)(7883) - (96)(1238)}{\sqrt{[(16)(808) - (96)^2][(16)(98196) - (1238)^2]}} = 0.61$$

$$r^2 = (0.61)^2 = 0.37$$

Yes, there is a linear association. In fact, there is a strong, positive correlation between hours of study and exam grade (r = +0.61); 37% of the variance in grades is explained by hours studied.

2. The necessary sums were provided in the previous problem.

$$b = \frac{N\,\Sigma XY - (\Sigma X)(\Sigma Y)}{N\,\Sigma X^2 - (\Sigma X)^2} = \frac{(16)(7883) - (96)(1238)}{(16)(808) - (96)^2} = 1.97$$

$$a = \overline{Y} - b\overline{X} = (77.375) - (1.85)(6.0) = 65.6$$

For every one hour that a person studies, we predict an average increase in his/her grade of 1.97 points. When a student doesn't study at all, the exam grade is predicted to be 65.6.

3.  The regression equation is: $\hat{Y} = 1.97X + 65.6$

Substituting in 4 for X: $\hat{Y} = 1.97(4) + 65.6 = 73.5$

Substituting in 6 for X: $\hat{Y} = 1.97(6) + 65.6 = 77.4$

Substituting in 8 for X: $\hat{Y} = 1.97(8) + 65.6 = 81.4$

Substituting in 10 for X: $\hat{Y} = 1.97(10) + 65.6 = 85.3$

Therefore, a student who studies for 4 hours is predicted to earn a 73.5 on the exam. A student who studies 6 hours is predicted to earn a 77.4. A student who studies 8 hours is predicted to earn an 81.4. A student who studies 10 hours is predicted to earn an 85.3.

4.  The necessary sums and quantities for calculating the slope (b), Y intercept (a), and Pearson's r are listed below.

| | | |
|---|---|---|
| $\Sigma X$ = 306 | $\Sigma Y$ = 805.6 | $\Sigma XY$ = 13703 |
| $\Sigma X^2$ = 4990 | $\Sigma Y^2$ = 56643.4 | N = 20 |
| $\overline{X}$ = 15.3 | $\overline{Y}$ = 40.28 | |

$$b = \frac{N\Sigma XY - (\Sigma X)(\Sigma Y)}{N\Sigma X^2 - (\Sigma X)^2} = \frac{(20)(13703) - (306)(805.6)}{(20)(4990) - (306)^2} = 4.5$$

$$a = \overline{Y} - b\overline{X} = (40.28) - (4.5)(15.3) = -28.1$$

$$r = \frac{N\Sigma XY - (\Sigma X)(\Sigma Y)}{\sqrt{[N\Sigma X^2 - (\Sigma X)^2][N\Sigma Y^2 - (\Sigma Y)^2]}}$$

$$= \frac{(20)(13703) - (306)(805.6)}{\sqrt{[(20)(4990) - (306)^2][(20)(56643.4) - (805.6)^2]}} = 0.50$$

$$r^2 = (0.50)^2 = 0.25$$

There is a moderate positive association between income and education. 25% of the variance in income can be explained by education. For each additional year of education, a person can expect an average increase in salary of $4,500. Based on this information, what do you think, should Matt go to grad school?

5.     The regression equation is: $\hat{Y} = 4.5X - 28.1$

Substituting in 12 for X: $\hat{Y} = 4.5(12) - 28.1 = 25.9$

Substituting in 16 for X: $\hat{Y} = 4.5(16) - 28.1 = 43.9$

Substituting in 18 for X: $\hat{Y} = 4.5(18) - 28.1 = 52.9$

Substituting in 22 for X: $\hat{Y} = 4.5(22) - 28.1 = 70.9$

Therefore, a person with a high school diploma is predicted to earn $25,900. A person with a college degree is predicted to earn $43,900. A person with a master's degree is predicted to earn $52,900. A person with a doctorate degree is predicted to earn $70,900.

6.     The necessary sums and quantities for calculating the slope (b), Y intercept (a), and Pearson's r are listed below.

$\Sigma X = 782$      $\Sigma Y = 50$      $\Sigma XY = 1735$

$\Sigma X^2 = 66752$      $\Sigma Y^2 = 200$      $N = 16$

$\overline{X} = 48.9$      $\overline{Y} = 3.1$

$$b = \frac{N\Sigma XY - (\Sigma X)(\Sigma Y)}{N\Sigma X^2 - (\Sigma X)^2} = \frac{(16)(1735) - (782)(50)}{(16)(66752) - (782)^2} = -0.02$$

$$a = \overline{Y} - b\overline{X} = (3.1) - (-0.02)(48.9) = 4.3$$

$$r = \frac{N\Sigma XY - (\Sigma X)(\Sigma Y)}{\sqrt{[N\Sigma X^2 - (\Sigma X)^2][N\Sigma Y^2 - (\Sigma Y)^2]}}$$

$$= \frac{(16)(1735) - (782)(50)}{\sqrt{[(16)(66752) - (782)^2][(16)(200) - (50)^2]}} = -0.63$$

$$r^2 = (-0.63)^2 = 0.40$$

Yes, there is a linear association. In fact, there is a negative, strong association between income and number of children. As income increases, the number of children a couple plans to have decreases; 40% of the variance in number of children is explained by income. For every one unit change in income, we can expect the number of children to decrease on average by 0.02.

7.  The necessary sums and quantities for calculating the slope (b), Y intercept (a), and Pearson's r are listed below.

$\Sigma X = 196 \qquad \Sigma Y = 298 \qquad \Sigma XY = 4236$

$\Sigma X^2 = 2830 \qquad \Sigma Y^2 = 6498 \qquad N = 14$

$\overline{X} = 14 \qquad \overline{Y} = 21.3$

$$b = \frac{N\Sigma XY - (\Sigma X)(\Sigma Y)}{N\Sigma X^2 - (\Sigma X)^2} = \frac{(14)(4236) - (196)(298)}{(14)(2830) - (196)^2} = 0.7$$

$$a = \overline{Y} - b\overline{X} = (21.3) - (0.7)(14) = 10.9$$

$$r = \frac{N\Sigma XY - (\Sigma X)(\Sigma Y)}{\sqrt{[N\Sigma X^2 - (\Sigma X)^2][N\Sigma Y^2 - (\Sigma Y)^2]}}$$

$$= \frac{(14)(4236) - (196)(298)}{\sqrt{[(14)(2830) - (196)^2][(14)(6498) - (298)^2]}} = 0.55$$

$$r^2 = (0.55)^2 = 0.31$$

There is a moderate positive association between education and age wed; 31% of the variance in age wed is explained by education. For every one year increase in education, we can expect the age of first married to increase on average by 0.7 years. To answer the sociologist, I would say "yes," education does lead to the postponement of marriage.

8.  The necessary sums and quantities for calculating the slope (b), Y intercept (a), and Pearson's r are listed below.

$\Sigma X = 822 \qquad \Sigma Y = 831.2 \qquad \Sigma XY = 45258.2$

$\Sigma X^2 = 38634 \qquad \Sigma Y^2 = 67693.6 \qquad N = 19$

$\overline{X} = 43.3 \qquad \overline{Y} = 43.7$

$$b = \frac{N\Sigma XY - (\Sigma X)(\Sigma Y)}{N\Sigma X^2 - (\Sigma X)^2} = \frac{(19)(45258.2) - (822)(831.2)}{(19)(38634) - (822)^2} = 3.02$$

$$a = \overline{Y} - b\overline{X} = (43.7) - (3.02)(43.3) = -87.2$$

$$r = \frac{N\Sigma XY - (\Sigma X)(\Sigma Y)}{\sqrt{[N\Sigma X^2 - (\Sigma X)^2][N\Sigma Y^2 - (\Sigma Y)^2]}}$$

$$= \frac{(19)(45258.2) - (822)(831.2)}{\sqrt{[(19)(38634) - (822)^2][(19)(67693.6) - (831.2)^2]}} = 0.95$$

$$r^2 = (0.95)^2 = 0.90$$

Yes, thee is a linear association; in fact, there is a very strong, positive association between hours worked per week and income, with 90% of the variance explained in income. For every additional hour worked, income is expected to increase on average by $3,020.

9. The necessary sums and quantities for calculating the slope (b), Y intercept (a), and Pearson's r are listed below.

| | | |
|---|---|---|
| $\Sigma X = 97$ | $\Sigma Y = 53$ | $\Sigma XY = 299$ |
| $\Sigma X^2 = 561$ | $\Sigma Y^2 = 185$ | $N = 20$ |
| $\overline{X} = 5.1$ | $\overline{Y} = 2.8$ | |

$$b = \frac{N\Sigma XY - (\Sigma X)(\Sigma Y)}{N\Sigma X^2 - (\Sigma X)^2} = \frac{(20)(299) - (97)(53)}{(20)(561) - (97)^2} = 0.43$$

$$a = \overline{Y} - b\overline{X} = (2.8) - (0.43)(5.1) = 0.58$$

$$r = \frac{N\Sigma XY - (\Sigma X)(\Sigma Y)}{\sqrt{[N\Sigma X^2 - (\Sigma X)^2][N\Sigma Y^2 - (\Sigma Y)^2]}}$$

$$= \frac{(20)(299) - (97)(53)}{\sqrt{[(20)(561) - (97)^2][(20)(185) - (53)^2]}} = 0.66$$

$$r^2 = (0.66)^2 = 0.44$$

There is a strong positive association between hours of television watched and number of messages seen; 44% of the variance in message recall is explained by television viewing. For every hour of TV watched, an average an additional 0.43 messages are seen. Yes, the data support the researcher's hypothesis.

10. The necessary sums and quantities for calculating the slope (b), Y intercept (a), and Pearson's r are listed below.

$$\Sigma X = 763 \qquad \Sigma Y = 251 \qquad \Sigma XY = 6837$$
$$\Sigma X^2 = 47985 \qquad \Sigma Y^2 = 6827 \qquad N = 19$$
$$\overline{X} = 40.2 \qquad \overline{Y} = 13.2$$

$$b = \frac{N\Sigma XY - (\Sigma X)(\Sigma Y)}{N\Sigma X^2 - (\Sigma X)^2} = \frac{(19)(6837) - (763)(251)}{(19)(47985) - (763)^2} = -0.19$$

$$a = \overline{Y} - b\overline{X} = (13.2) - (-0.19)(40.2) = 20.7$$

$$r = \frac{N \cdot \Sigma XY - (\Sigma X)(\Sigma Y)}{\sqrt{[N\Sigma X^2 - (\Sigma X)^2][N\Sigma Y^2 - (\Sigma Y)^2]}}$$

$$= \frac{(19)(6837) - (763)(251)}{\sqrt{[(19)(47985) - (763)^2][(19)(6827) - (251)^2]}} = -0.42$$

$$r^2 = (-0.42)^2 = 0.17$$

There is a moderate, negative relationship between income and religious services. Only 18% of the variance in number of religious services attended is explained by income, though. In sum, religiosity does decrease with income; people with higher income attend fewer religious services.

11. The necessary sums and quantities for calculating the slope (b), Y intercept (a), and Pearson's r are listed below.

$$\Sigma X = 28 \qquad \Sigma Y = 192 \qquad \Sigma XY = 436$$
$$\Sigma X^2 = 96 \qquad \Sigma Y^2 = 2644 \qquad N = 15$$
$$\overline{X} = 1.9 \qquad \overline{Y} = 12.8$$

$$b = \frac{N\Sigma XY - (\Sigma X)(\Sigma Y)}{N\Sigma X^2 - (\Sigma X)^2} = \frac{(15)(436) - (28)(192)}{(15)(96) - (28)^2} = 1.8$$

$$a = \overline{Y} - b\overline{X} = (12.8) - (1.8)(1.9) = 9.5$$

$$r = \frac{N \Sigma XY - (\Sigma X)(\Sigma Y)}{\sqrt{[N \Sigma X^2 - (\Sigma X)^2][N \Sigma Y^2 - (\Sigma Y)^2]}}$$

$$= \frac{(15)(436) - (28)(192)}{\sqrt{[(15)(96) - (28)^2][(15)(2644) - (192)^2]}} = 0.86$$

$$r^2 = (0.86)^2 = 0.74$$

There is a strong, positive association between number of children and hours of study per week; 74% of the variance in hours studied is explained by the number of children. As the number of children increases, so too do the hours of study. In fact, for every extra child, the professor can expect the student to study an average of 1.8 hours more per week.

12.  The necessary sums and quantities for calculating the slope (b), Y intercept (a), and Pearson's r are listed below.

| | | |
|---|---|---|
| $\Sigma X = 299$ | $\Sigma Y = 110$ | $\Sigma XY = 1440$ |
| $\Sigma X^2 = 7523$ | $\Sigma Y^2 = 1802$ | $N = 15$ |
| $\overline{X} = 19.9$ | $\overline{Y} = 7.3$ | |

$$b = \frac{N \Sigma XY - (\Sigma X)(\Sigma Y)}{N \Sigma X^2 - (\Sigma X)^2} = \frac{(15)(1440) - (299)(110)}{(15)(7523) - (299)^2} = -0.48$$

$$a = \overline{Y} - b\overline{X} = (7.3) - (-0.48)(19.9) = 16.9$$

$$r = \frac{N \Sigma XY - (\Sigma X)(\Sigma Y)}{\sqrt{[N \Sigma X^2 - (\Sigma X)^2][N \Sigma Y^2 - (\Sigma Y)^2]}}$$

$$= \frac{(15)(1440) - (299)(110)}{\sqrt{[(15)(7523) - (299)^2][(15)(1802) - (110)^2]}} = -0.60$$

$$r^2 = (-0.60)^2 = 0.36$$

Yes, there is a negative association between these two variables, which indicates that the closer one lives to the city center, the more events s/he will attend; 36% of the variance in number of events is explained by proximity to city. For each additional mile away from the city center a person lives, we can predict him/her to attend on average 0.48 fewer events.

13.  The necessary sums and quantities for calculating the slope (b), Y intercept (a), and Pearson's r are listed below.

$$\Sigma X = 139 \qquad \Sigma Y = 148 \qquad \Sigma XY = 1057$$
$$\Sigma X^2 = 1655 \qquad \Sigma Y^2 = 3460 \qquad N = 14$$
$$\overline{X} = 9.9 \qquad \overline{Y} = 10.6$$

$$b = \frac{N\Sigma XY - (\Sigma X)(\Sigma Y)}{N\Sigma X^2 - (\Sigma X)^2} = \frac{(14)(1057) - (139)(148)}{(14)(1655) - (139)^2} = -1.5$$

$$a = \overline{Y} - b\overline{X} = (10.6) - (-1.5)(9.9) = 25.5$$

$$r = \frac{N\Sigma XY - (\Sigma X)(\Sigma Y)}{\sqrt{[N\Sigma X^2 - (\Sigma X)^2][N\Sigma Y^2 - (\Sigma Y)^2]}}$$

$$= \frac{(14)(1057) - (139)(148)}{\sqrt{[(14)(1655) - (139)^2][(14)(3460) - (148)^2]}} = -0.57$$

$$r^2 = (-0.57)^2 = 0.33$$

There is a moderate negative association between age of child and number of family events. The data do support the hypothesis that as children grow older, families spend less time together.

14.  The necessary sums and quantities for calculating the slope (b), Y intercept (a), and Pearson's r are listed below.

$$\Sigma X = 371 \qquad \Sigma Y = 116 \qquad \Sigma XY = 1202$$
$$\Sigma X^2 = 8293 \qquad \Sigma Y^2 = 976 \qquad N = 23$$
$$\overline{X} = 16.1 \qquad \overline{Y} = 5.0$$

$$b = \frac{N\Sigma XY - (\Sigma X)(\Sigma Y)}{N\Sigma X^2 - (\Sigma X)^2} = \frac{(23)(1202) - (371)(116)}{(23)(8293) - (371)^2} = -0.29$$

$$a = \overline{Y} - b\overline{X} = (5.0) - (-0.29)(16.1) = 9.7$$

$$r = \frac{N\,\Sigma XY - (\Sigma X)(\Sigma Y)}{\sqrt{[N\,\Sigma X^2 - (\Sigma X)^2][N\,\Sigma Y^2 - (\Sigma Y)^2]}}$$

$$= \frac{(23)(1202) - (371)(116)}{\sqrt{[(23)(8293) - (371)^2][(23)(976) - (116)^2]}} = -0.69$$

$$r^2 = (-0.69)^2 = 0.48$$

There is a fairly strong, negative association between hours of free time and number of activities, which indicates that people who are busier do get more done. 49% of the variance in number of activities is explained by hours of free time per day.

15. The necessary sums and quantities for calculating the slope (b), Y intercept (a), and Pearson's r are listed below.

| $\Sigma X$ | = 70 | $\Sigma Y$ | = 40 | $\Sigma XY$ | = 208 |
|---|---|---|---|---|---|
| $\Sigma X^2$ | = 376 | $\Sigma Y^2$ | = 166 | N | = 16 |
| $\overline{X}$ | = 4.4 | $\overline{Y}$ | = 2.5 | | |

$$b = \frac{N\,\Sigma XY - (\Sigma X)(\Sigma Y)}{N\,\Sigma X^2 - (\Sigma X)^2} = \frac{(16)(208) - (70)(40)}{(16)(376) - (70)^2} = 0.47$$

$$a = \overline{Y} - b\overline{X} = (2.5) - (0.47)(4.4) = 0.43$$

$$r = \frac{N\,\Sigma XY - (\Sigma X)(\Sigma Y)}{\sqrt{[N\,\Sigma X^2 - (\Sigma X)^2][N\,\Sigma Y^2 - (\Sigma Y)^2]}}$$

$$= \frac{(16)(208) - (70)(40)}{\sqrt{[(16)(376) - (70)^2][(16)(166) - (40)^2]}} = 0.49$$

$$r^2 = (0.49)^2 = 0.24$$

Yes, there is a linear association. In fact, there is a moderate positive association between television viewing and violence among children, with 24% of the variance in violence explained by hours of television watching. For every hour of TV viewing, we predict an average increase of 0.47 in violent acts.

## SPSS or MicroCase WORK PROBLEMS

If you had problems with any of these questions, be sure to re-read the appropriate sections at the end of the chapter in your textbook.

1.    To save space, I ran all of the correlations at once.

**Correlations**

| | | SIBS | CHILDS | REALINC | AGE | TVHOURS |
|---|---|---|---|---|---|---|
| SIBS | Pearson Correlation | 1 | .219(**) | -.142(**) | .095(**) | .138(**) |
| | Sig. (2-tailed) | . | .000 | .000 | .003 | .000 |
| | N | 1002 | 1002 | 855 | 1000 | 672 |
| CHILDS | Pearson Correlation | .219(**) | 1 | -.005 | .399(**) | .071 |
| | Sig. (2-tailed) | .000 | . | .864 | .000 | .065 |
| | N | 1002 | 1497 | 1281 | 1490 | 672 |
| REALINC | Pearson Correlation | -.142(**) | -.005 | 1 | -.012 | -.255(**) |
| | Sig. (2-tailed) | .000 | .864 | . | .655 | .000 |
| | N | 855 | 1281 | 1282 | 1278 | 573 |
| AGE | Pearson Correlation | .095(**) | .399(**) | -.012 | 1 | .128(**) |
| | Sig. (2-tailed) | .003 | .000 | .655 | . | .001 |
| | N | 1000 | 1490 | 1278 | 1493 | 672 |
| TVHOURS | Pearson Correlation | .138(**) | .071 | -.255(**) | .128(**) | 1 |
| | Sig. (2-tailed) | .000 | .065 | .000 | .001 | . |
| | N | 672 | 672 | 573 | 672 | 673 |

** Correlation is significant at the 0.01 level (2-tailed).

There is a weak (almost moderate) correlation between the number of siblings and the number of children one has (r=+0.219). As the number of siblings increases, the number of offspring increases.

To predict values of Y given values of X, we need to use the regression command. Click on the "Analyze" menu; click on the "Regression" option; and then click on "Linear." The dependent variable is *childs*, and the independent is *sibs*. After clicking on the "OK" button, the following output should appear. In the second box, you have the coefficient of determination ($R^2$), which is 0.047, meaning that the 4.7% of variation in the number of children one has is predicted by the number of siblings one has.

**Model Summary**

| Model | R | R Square | Adjusted R Square | Std. Error of the Estimate |
|---|---|---|---|---|
| 1 | .219 | .048 | .047 | 1.597 |

**Coefficients(a)**

| Model | Unstandardized Coefficients | | Standardized Coefficients | t | Sig. |
|---|---|---|---|---|---|
| | B | Std. Error | Beta | | |
| 1  (Constant) | 1.437 | .080 | | 18.004 | .000 |
| SIBS | .116 | .016 | .219 | 7.098 | .000 |

Next, to make our predictions, we need to plug our X values into the regression equation. The computer gives us a (the Y intercept) and b (the slope), so all we need to do is solve for Y by plugging X into the equation:

$$Y = a + bX = (1.437) + (0.116)(2) = 1.669$$

Thus, we would expect people with 2 siblings to have on average 1.7 children.

For a person with 10 siblings, we would simply plug in 10 instead of 2 for the value of X:

$$Y = a + bX = (1.437) + (0.116)(10) = 2.597$$

Thus, we would expect people with 10 siblings to have on average 2.6 children.

2.  There is a very weak negative relationship between income and number of children ($r = -0.005$), indicating that as income increases, the number of children one has decreases. If we square this number, we see that 0.003% ($r^2 = 0.000025$) of the variance in number of children is accounted for by the number of siblings. Therefore, the hypothesis is supported.

3.  There is a moderate positive association between number of children and age ($r = 0.399$), indicating that as age increases, the number of children one has increases. Age explains 16% of the variance in the number of offspring ($r^2 = 0.16$).

4.  There is a very weak negative association between income and age ($r = -0.012$), indicating that as age increases, income decreases. Age explains .01% of the variance in income ($r^2 = 0.0001$).

5.  There is a moderate negative association between income and TV viewership ($r = -.255$), indicating that as income increases, viewership decreases. Income and TV viewership share 6.5 percent of their variation ($r^2 = 0.0650$).

# CHAPTER FIFTEEN
## Partial Correlation and Multiple Regression and Correlation

*Learning Objectives*: *By the end of this chapter, you will be able to:*

1. Compute and interpret partial correlation coefficients.
2. Find and interpret the least-squares multiple regression equation with partial slopes.
3. Calculate and interpret the multiple correlation coefficient ($R^2$).
4. Explain the limitations of partial and multiple regression analysis.


## KEY TERMS

- ✓ Beta-Weights ($b^*$)  (p. 371)
- ✓ Coefficient of Multiple Determination ($R^2$)  (p. 373)
- ✓ Multiple Correlation  (p. 373)
- ✓ Multiple Correlation Coefficient ($R$)  (p. 373)
- ✓ Multiple Regression  (p. 367)
- ✓ Partial Correlation  (p. 363)
- ✓ Partial Correlation Coefficient  (p. 364)
- ✓ Partial Slopes  (p. 367)
- ✓ Standardized Partial Slopes (beta-weights)  (p. 371)
- ✓ Zero-Order Correlations  (p. 364)


## CHAPTER SUMMARY

In this chapter multivariate techniques for interval-ratio variables are introduced.  Specifically, partial correlation, multiple regression, and multiple correlation are discussed.


## CHAPTER OUTLINE

### 15.1  Introduction.

### 15.2  Partial Correlation.

- In partial correlation analysis the relationship between X and Y is analyzed, then the effects of a third variable, Z, are controlled.  These third variables are called control variables.  When we calculated the correlation between two variables, we were computing Pearson's r.  These correlations are also called zero-order correlations and refer to the correlation between two variables.  In order to control for the effects of the third variable, Z, we need to use partial correlation analysis.  Partial correlations help us examine the relationship between X and Y controlling for Z (symbolically given as $r_{yx.z}$).

- The zero-order correlation ($r_{yx}$) is then compared to the partial correlation ($r_{yx.z}$). There are three patterns we are concerned with. First, if the zero-order correlation is approximately equal to the partial correlation, then we conclude that the control variable has little or no effect on the bivariate relationship because the correlation didn't change once we controlled for Z. This is called a direct relationship between X and Y.
- The second possible pattern occurs when the zero-order correlation is much larger than the partial correlation ($r_{yx} > r_{yx.z}$). We call this pattern either spurious or intervening, meaning that the relationship between X and Y drops off once the third variable is controlled. Deciding whether the relationship is spurious or intervening must be based on theory or the logical ordering of the variables. When Z is the cause of both X and Y (that is, Z logically occurs before X and Y), we have a spurious relationship. X and Y only appeared to be related (according to the bivariate, or zero-order, correlation) because they shared a common cause, Z. When Z causally links X and Y (that is, Z logically occurs between X and Y), we have an intervening relationship.
- The third possible pattern cannot be detected by partial correlation analysis, but because it leads to very interesting conclusions, it deserves a brief introduction. This third pattern is called interaction. Interaction occurs when the relationship between X and Y changes for different values of Z. For example, let's say we control for gender (Z) while examining the relationship between marital status (X) and income (Y). Now let's say that we find that the correlation between marital status and income is negative for women (meaning that married women earn less than single women) and positive for men (meaning that married men earn more than single men). In this instance gender and marital status interact in their effects on income.

## 15.3 Multiple Regression: Predicting the Dependent Variable.

- Let's turn now to multiple regression, a statistical technique for a single interval-ratio level dependent variable and two or more interval-ratio independent variables. There are two main parts to multiple regression. First, as we saw in least squares regression, we can make predictions of the dependent variable with the regression equation. When we introduce additional independent variables, the regression equation changes to reflect this:

$$Y = a + b_1 X_1 + b_2 X_2$$

- Where a is the Y intercept; $X_1$ is a score on the first independent variable; $X_2$ is a score on the second independent variable; $b_1$ is the partial slope for the relationship between $X_1$ and Y; and $b_2$ is the partial slope for the relationship between $X_2$ and Y.
- Or, in the case of more than two independent variables, the regression equation becomes:

$$Y = a + b_1 X_1 + b_2 X_2 + b_3 X_3 + \ldots + b_k X_k$$

- In this equation, we need to calculate a (the Y-intercept) and $b_1$ and $b_2$ (the partial slopes for each of the independent variables). Partial slopes indicate the change in the dependent variable for a one unit change in the independent variable, controlling for all of the other independent variables in the equation. We can then plug the values of $X_1$ and $X_2$ into the equation, as we did in chapter 13, and make a prediction for the value of Y.

## 15.4 Multiple Regression: Assessing the Effects of the Independent Variables.

- The second part to multiple regression is assessing the effects of the two independent variables: which variable X (now called $X_1$) or Z (now called $X_2$) has the largest effect on Y? In order to determine this, we need to convert all variables into Z scores.
- Think way back to chapter five. Remember how we converted variables into Z scores in order to change the scales? For example, temperatures measured on the Fahrenheit scale can be converted to the Celsius scale; the temperature hasn't changed, only the scale with which we're measuring it. Rather than a freezing point of $32°$, the new scale has a freezing point of $0°$. This is the same logic we are using when we want to assess the effects of $X_1$ and $X_2$ on Y. Because the variables are usually measuring different things (such as hours worked per week and education), it is difficult to compare their relative effects (the old "apples and oranges" problem). Instead, we could convert all the data to Z scores, such that each distribution has a mean of zero and a standard deviation of one, and then compare them. The simplest way to do this is to use the following equation:

$$Z_y = b_1^* Z_1 + b_2^* Z_2 \quad \text{or} \quad Z_y = b_1^* Z_1 + b_2^* Z_2 + \ldots\ldots + b_k^* Z_k$$

- Where $b_1^*$ and $b_2^*$ are the standardized partial slopes for the independent variables, or what are called beta weights. With these beta weights, you can assess the effect of each independent variable on the dependent variable separately to determine which has the strongest effect. Beta weights can range from -1 to +1; thus if $b_1^* = -0.76$ and $b_2^* = +0.66$, then $X_1$ has the strongest relative effect ($|-0.76| > |+0.66|$, the signs are inconsequential).
- Note that the standardized regression equation no longer has a Y intercept term ($a_z$). Because the standardized variables have means of zero, the Y intercept is also zero. This means that standardized regression lines will always pass through the origin.

## 15.5  Multiple Correlation.

- Multiple correlation techniques allow us to examine the effects of ALL the independent variables together by using the multiple correlation coefficient (R) and the coefficient of multiple determination ($R^2$).  When we were analyzing the association between two interval-ratio variables, we used Pearson's r and $r^2$, or the coefficient of determination.  The coefficient of determination ($r^2$) is a PRE-type measure which tells us the amount of variation explained in the dependent variable by the independent variable.  Similarly, the coefficient of *multiple* determination ($R^2$) is the amount of variation explained in the dependent variable by the first independent variable ($X_1$) and by the second independent variable ($X_2$) after the shared variation has been removed.  More succinctly, $R^2$ is the amount of variation explained in the dependent variable by all of the independent variables.  Thus, if we have an $R^2$ of 0.25, then we conclude that 25% of the variation (or variance) in the dependent variable is explained by the independent variables.

## 15.6  The Limitations of Multiple Regression and Correlation.

- As your textbook says, multiple regression and correlation are powerful statistical tools.  Indeed, multiple regression (sometimes called multivariate regression or ordinary least squares regression) is probably the most commonly used statistical technique in social science research.  But, as Healey says, powerful tools aren't cheap; that is, they require that certain assumptions be met before we can trust their results.  First, these techniques assume that each independent variable shares a linear relationship with the dependent variable; we can use scattergrams to see if we meet this assumption.  Second, the effects of all of the independent variables are assumed to be additive, not interactive or multiplicative.  Third, the independent variables must be uncorrelated with each other.  We rarely meet this particular assumption exactly; instead we consider this assumption met as long as the correlations among the independent variables are low.

## MULTIPLE-CHOICE QUESTIONS

1.  If we want to control for the effects of a third interval-ratio variable, we use
    _____.
    a. partial tables
    b. partial correlations
    c. multiple regression
    d. beta weights

    *Learning Objective: 1*

2. Multiple regression is used when you have _____ dependent variable(s) and _____ independent variable(s).
   a. two, two
   b. one, two
   c. one, three
   d. one, more than two

   *Learning Objective: 2*


3. If we want to make a prediction of the value of Y with more than one independent variable, we use _____.
   a. partial tables
   b. partial correlations
   c. multiple regression
   d. beta weights

   *Learning Objective: 2*


4. Beta weights are used to _____.
   a. make predictions
   b. determine the amount of variance explained in the dependent variable by the independent variables
   c. determine the effects of each of the independent variables on the dependent variable
   d. determine causality

   *Learning Objective: 2*


5. Partial slopes are _____.
   a. the slope of the relationship between one independent variable and the dependent variable, controlling for the effects of all other independent variables
   b. the amount of variance explained in the dependent variable by the independent variables
   c. the correlation for each partial table
   d. the combined effects of all the independent variables on the dependent variable

   *Learning Objective: 2*

6. The coefficient of multiple determination is _____.
   a. $r^2$
   b. $b^*$
   c. $R^2$
   d. $Z_y$

   *Learning Objective: 3*


7. The coefficient of multiple determination is the _____.
   a. slope of the relationship between a particular independent variable and the dependent variable
   b. amount of variation explained in the dependent variable by all independent variables combined
   c. combined effects of all the independent variables on the dependent variable
   d. slope of the relationship between one independent variable and the dependent variable, controlling for the effects of all other independent variables

   *Learning Objective: 3*


8. The symbol for the relationship between X and Y controlling for Z is _____.
   a. $R^2$
   b. $b_1^*$
   c. $r_{yx.z}$
   d. $r_{zx.y}$

   *Learning Objective: 1*


9. If we have two partial slopes, the first -0.26, the second 0.19, we can conclude _____.
   a. the second variable has a stronger effect
   b. there is no association between these two variables and the dependent variable
   c. the first variable has a stronger effect
   d. there is a mixed correlation between the variables

   *Learning Objective: 2*

10. The multiple correlation coefficient is used to _____.
   a. determine the slope of the relationship between a particular independent variable and the dependent variable
   b. determine the strength of the association between a dependent variable and two or more independent variables
   c. determine the amount of variance explained in the dependent variable by the independent variables
   d. determine the effects of each of the independent variables on the dependent variable

   *Learning Objective: 3*


11. In the standardized least-squares equation, _____ are used to normalize the distributions.
   a. partial slopes
   b. partial correlations
   c. Z scores
   d. all of these choices

   *Learning Objective: 2*


12. Zero-order correlations are used to indicate the correlation between _____.
   a. two variables
   b. ordinal variables
   c. one ordinal and one interval-ratio variable
   d. beta weights

   *Learning Objective: 1*


13. Unstandardized partial slopes indicate _____.
   a. the effect of a particular independent variable on the dependent variable
   b. the amount of variance explained in the dependent variable
   c. the predicted value of Z
   d. the amount change in Y for a unit change in X

   *Learning Objective: 2*

14. $Z_1$ is _____.
   a. the Z score for the first dependent variable
   b. the critical value for the amount of variance explained in the dependent variable
   c. the standardized score for the first independent variable
   d. the standardized score for the second independent variable

   *Learning Objective: 2*


15. The term $Z_y$ is used to denote _____.
   a. the critical value for the amount of variance explained in the dependent variable
   b. the standardized regression equation
   c. the amount of variance explained in the dependent variable
   d. the beta weights

   *Learning Objective: 2*


16. Which of the below is not one of the assumptions of multiple regression?
   a. the independent variables share linear relationships with the dependent variable
   b. the independent variables share multiplicative relationships with the dependent variable
   c. the independent variables are unrelated to each other
   d. the independent variables share additive relationships with the dependent variable

   *Learning Objective: 4*


17. Suppose that $r_{yx} = +.35$ and $r_{yx \bullet z} = +.36$ (and that Y is income, X is education, and Z is sex). These results indicate a _____ relationship.
   a. spurious
   b. intervening
   c. direct
   d. interaction

   *Learning Objective: 1*


18. Suppose that $r_{yx} = +.35$ and $r_{yx \bullet z} = +.02$ (and that Y is income, X is occupational prestige, and Z is age). These results indicate a _____ relationship.
   a. spurious
   b. intervening
   c. direct
   d. interaction

   *Learning Objective: 1*

19. Suppose that $r_{yx} = +.35$ and $r_{yx \bullet z} = -.01$ (and that Y is income, X is sex, and Z is hours worked per week). These results indicate a _____ relationship.
   a. spurious
   b. intervening
   c. direct
   d. interaction

*Learning Objective: 1*


20. Suppose the correlation between occupational prestige and education is +.65 for men and -.15 for women. These results indicate a _____ relationship.
   a. spurious
   b. intervening
   c. direct
   d. interaction

*Learning Objective: 1*


## WORK PROBLEMS

1. A researcher was interested in the relationship between education and income, and although they had a fairly strong, positive relationship, she thought that job prestige might explain more of the variance. Below are the correlations, means, and standard deviations for the data. Determine the regression equation.

|  | EDUC ($X_1$) | PRESTIGE ($X_2$) | INCOME (Y) |
|---|---|---|---|
| EDUC ($X_1$) | 1.0 | 0.46 | 0.54 |
| PRESTIGE ($X_2$) |  | 1.0 | 0.32 |
| INCOME (Y) |  |  | 1.0 |
| MEAN $(\overline{X})$ | 13.5 | 42.1 | $32,590 |
| STANDARD DEVIATION (s) | 3.7 | 15.6 | 8,970 |

*Learning Objective: 2*

2. Using the information from problem 1, predict the income for someone who has 15 years of education and a prestige score of 63.

*Learning Objective: 2*

3. Using the information from problem 1, verify the parital slope for education by predicting the income for someone who has 16 years of education and a prestige score of 63.

*Learning Objective: 2*

4. Using the information from problem 1, verify the parital slope for prestige by predicting the income for someone who has 15 years of education and a prestige score of 64.

*Learning Objective: 2*

5. Using the information from problem 1, which independent variable has the strongest effect on the dependent variable?

*Learning Objective: 2*

6. Using the information from problem 1, how much variance is explained in the dependent variable by the independent variable?

*Learning Objective: 3*

7. Quality of life of AIDS patients was hypothesized to be affected by health status and the length of time a person was HIV positive. The data are presented below. Which independent variable has the strongest effect on the dependent variable?

| | HEALTH $(X_1)$ | TIME HIV+ $(X_2)$ | QUALITY OF LIFE (Y) |
|---|---|---|---|
| HEALTH $(X_1)$ | 1.0 | -0.31 | 0.63 |
| TIME HIV+ $(X_2)$ | | 1.0 | -0.43 |
| QUALITY OF LIFE (Y) | | | 1.0 |
| MEAN $\left(\overline{X}\right)$ | 10.8 | 53.3 | 42.3 |
| STANDARD DEVIATION (s) | 6.1 | 34.1 | 29.6 |

*Learning Objective: 2*

8. Using the information from problem 7, how much variance is explained in the dependent variable by the independent variable?

*Learning Objective: 3*

9. In a study of knowledge of family planning, a researcher found that two variables were correlated with knowledge: hours of television watched per day and cosmopolitanism (worldliness). Discuss each of the relationships in the correlation matrix.

| | TV/DAY $(X_1)$ | COSMO $(X_2)$ | FAM. PLAN. (Y) |
|---|---|---|---|
| TV HOURS A DAY $(X_1)$ | 1.0 | 0.05 | -0.47 |
| COSMOPOLITANISM $(X_2)$ | | 1.0 | 0.56 |
| FAMILY PLANNING (Y) | | | 1.0 |
| MEAN $\left(\overline{X}\right)$ | 4.3 | 27.6 | 8.3 |
| STANDARD DEVIATION (s) | 1.5 | 18.3 | 2.9 |

*Learning Objective: 1 & 2*

10. Using the information from problem 9, which variable has the strongest effect on the dependent variable?

   *Learning Objective: 1*

11. How much variance in knowledge of family planning is explained by these variables?

   *Learning Objective: 3*

12. The Wesaloosa dorm guys are still trying to figure out why they aren't getting any dates. They decided to find out what might be the primary reason. Given the data below, what would you conclude?

| | HRS OF STUDY $(X_1)$ | CLEANLINESS $(X_2)$ | DATES $(Y)$ |
|---|---|---|---|
| HOURS OF STUDY $(X_1)$ | 1.0 | 0.09 | -0.59 |
| CLEANLINESS $(X_2)$ | | 1.0 | -0.37 |
| NUMBER OF DATES $(Y)$ | | | 1.0 |
| MEAN $\left(\overline{X}\right)$ | 3.5 | 56 | 2.1 |
| STANDARD DEVIATION $(s)$ | 1.3 | 23 | 1.6 |

   *Learning Objective: 1 & 2*

13. A criminologist was interested in the relationship between truancy and criminal acts committed by teenagers. He hypothesized that socioeconomic status might also have an effect. Which variable has the strongest effect?

| | TRUANCY $(X_1)$ | SES $(X_2)$ | CRIMINAL ACTS $(Y)$ |
|---|---|---|---|
| TRUANCY $(X_1)$ | 1.0 | -0.35 | 0.68 |
| SES $(X_2)$ | | 1.0 | -0.22 |
| CRIMINAL ACTS $(Y)$ | | | 1.0 |
| MEAN $\left(\overline{X}\right)$ | 16 | 46.2 | 3.1 |
| STANDARD DEVIATION $(s)$ | 23 | 24.3 | 2.3 |

   *Learning Objective: 1 & 2*

14. How much variance is explained in criminal activity by these two variables?

*Learning Objective: 3*

## SPSS or MicroCase WORK PROBLEMS

The following work problems can be completed using either SPSS or MicroCase. The solutions will be the same regardless of which one you use. Your instructor may have a preference, so be sure to check with him/her prior to beginning these problems.

1. The variables age *(age)* and the number of siblings *(sibs)* are associated with number of children *(childs)* one has. Which of these variables has the strongest effect?

2. It is hypothesized that age *(age)*, income *(realinc)*, prestige *(sei)*, and the number of children one has *(childs)* affect the number of hours of television *(tvhours)* a person watches per day. Which of these variables has the *strongest* effect on TV hours?

3. The variables occupational prestige *(sei)*, the number of children *(childs)*, and age *(age)* are associated with income *(realinc)*. Which of these variables has the strongest effect?

4. The variables number of siblings *(sibs)*, television viewership *(tvhours)*, and age *(age)* are associated with occupational prestige *(sei)*. Which of these variables has the strongest effect?

## ANSWER KEY

## MULTIPLE-CHOICE QUESTIONS

| | | | | | | |
|---|---|---|---|---|---|---|
| 1. | b | (p. 362) | 11. | c | (p. 371) |
| 2. | d | (p. 367) | 12. | a | (p. 364) |
| 3. | c | (p. 367) | 13. | d | (p. 367) |
| 4. | c | (p. 371) | 14. | c | (p. 373) |
| 5. | a | (p. 367-368) | 15. | b | (p. 373) |
| 6. | d | (p. 373-374) | 16. | b | (p. 375-379) |
| 7. | c | (p. 373-374) | 17. | c | (p. 363) |
| 8. | c | (p. 364-365) | 18. | a | (p. 363-364) |
| 9. | c | (p. 367-368) | 19. | b | (p. 363-364) |
| 10. | b | (p. 373-374) | 20. | d | (p. 364) |

## WORK PROBLEMS

1. In order to calculate predicted Y, we first have to calculate $b_1$, $b_2$, and a from the information provided.

$$b_1 = \left(\frac{s_y}{s_1}\right)\left(\frac{r_{y1} - r_{y2}r_{12}}{1 - r_{12}^2}\right) = \left(\frac{8970}{3.7}\right)\left(\frac{(0.54) - (0.32)(0.46)}{1 - 0.46^2}\right) = 1{,}207.86$$

$$b_2 = \left(\frac{s_y}{s_2}\right)\left(\frac{r_{y2} - r_{y1}r_{12}}{1 - r_{12}^2}\right) = \left(\frac{8970}{15.6}\right)\left(\frac{(0.32) - (0.54)(0.46)}{1 - 0.46^2}\right) = 52.22$$

$$a = \overline{Y} - b_1\overline{X}_1 - b_2\overline{X}_2 = (32950) - (1207.86)(13.5) - (52.22)(42.1) = 14{,}445.43$$

The regression equation is: $Y = 14445.43 + 1207.86X_1 + 52.22X_2$

2. The regression equation is: $Y = 14445.43 + 1207.86X_1 + 52.22X_2$

Substituting the given values: $Y = 14445.43 + 1207.86(15) + 52.22(63) = 35{,}853.19$

We predict that a person with 15 years of education and job with a prestige score of 63 will make approximately \$35,853 per year.

3. The regression equation is: $Y = 14445.43 + 1207.86X_1 + 52.22X_2$

Substituting the given values: $Y = 14445.43 + 1207.86(16) + 52.22(63) = 37{,}061.05$

We had predicted that someone with 15 years of education and 63 prestige points would earn \$35,853.19. Subtracting this value from the one we just calculated above, 37061.05-35853.19, equals 1207.86, which verifies the regression coefficient for education.

4. The regression equation is: $Y = 14445.43 + 1207.86X_1 + 52.22X_2$

Substituting the given values: $Y = 14445.43 + 1207.86(15) + 52.22(64) = 35{,}905.41$

We had predicted that someone with 15 years of education and 63 prestige points would earn \$35,853.19. Subtracting this value from the one we just calculated above, 35905.41-35853.19, equals 52.22, which verifies the regression coefficient for prestige.

5. To determine which independent variable has the strongest effect we have to calculate the standardized partial slopes ($b^*$).

$$b_1^* = b_1 \left( \frac{s_1}{s_y} \right) = (1207.86)\left( \frac{3.7}{8970} \right) = 0.50$$

$$b_2^* = b_2 \left( \frac{s_2}{s_y} \right) = (52.22)\left( \frac{15.6}{8970} \right) = 0.09$$

Therefore, the first independent variable, education, has the strongest effect on the dependent variable, income.

6. In order to determine the amount of variance in the dependent variable that is explained by the independent variable, we have to calculate $R^2$.

$$r_{y2 \bullet 1} = \frac{r_{y2} - (r_{y1})(r_{12})}{\sqrt{1 - r_{y1}^2}\sqrt{1 - r_{12}^2}} = \frac{(0.32) - (0.54)(0.46)}{\sqrt{1 - (0.54)^2}\sqrt{1 - (0.46)^2}} = 0.10$$

$$R^2 = r_{y1}^2 + r_{y2 \bullet 1}^2 (1 - r_{y1}^2) = (0.54)^2 + (0.10)^2[1 - (0.54)^2] = 0.30$$

Therefore, 30% of the variance in income is explained by education and occupational prestige.

7. In order to determine which variable has the strongest effect on the dependent variable we have to calculate that standardized partial slopes. But, to calculate the standardized partial slopes, we first have to calculate the (unstandardized) partial slopes.

$$b_1 = \left( \frac{s_y}{s_1} \right)\left( \frac{r_{y1} - r_{y2}r_{12}}{1 - r_{12}^2} \right) = \left( \frac{29.6}{6.1} \right)\left( \frac{(0.63) - (-0.43)(-0.31)}{1 - (-0.31)^2} \right) = 2.67$$

$$b_2 = \left( \frac{s_y}{s_2} \right)\left( \frac{r_{y2} - r_{y1}r_{12}}{1 - r_{12}^2} \right) = \left( \frac{29.6}{34.1} \right)\left( \frac{(-0.43) - (0.63)(-0.31)}{1 - (-0.31)^2} \right) = -0.23$$

$$b_1^* = b_1 \left( \frac{s_1}{s_y} \right) = (2.67)\left( \frac{6.1}{29.6} \right) = 0.55$$

$$b_2^* = b_2 \left( \frac{s_2}{s_y} \right) = (-0.23)\left( \frac{34.1}{29.6} \right) = -0.26$$

Health status appears to have the stronger effect on quality of life. Note that health is positive, indicating that healthier individuals have higher quality of life, while length of time is negative, indicating that the longer one is HIV positive, the lower his/her quality of life.

8. In order to determine the amount of variance in the dependent variable that is explained by the independent variable, we have to calculate $R^2$.

$$r_{y2\bullet1} = \frac{r_{y2} - (r_{y1})(r_{12})}{\sqrt{1-r_{y1}^2}\sqrt{1-r_{12}^2}} = \frac{(-0.43) - (0.63)(-0.31)}{\sqrt{1-(0.63)^2}\sqrt{1-(-0.31)^2}} = -0.32$$

$$R^2 = r_{y1}^2 + r_{y2\bullet1}^2(1-r_{y1}^2) = (0.63)^2 + (-0.32)^2[1-(0.63)^2] = 0.46$$

Therefore, 46% of the variance in Quality of Life is explained by health and length of time HIV positive.

9. Hours of TV watching per day has a moderate negative relationship with knowledge of family planning (r=-0.47), indicating that as TV viewing increases, knowledge of family planning decreases. Cosmopolitanism and knowledge of family planning have a moderate positive association (r=0.56). Therefore, as cosmopolitanism increases, knowledge of family planning also increases. Finally, cosmopolitanism and TV watching share a very weak, positive relationship (r=0.05); cosmopolitanism increases slightly with TV viewing.

10. In order to determine which variable has the strongest effect on the dependent variable we have to calculate that standardized partial slopes. But, to calculate the standardized partial slopes, we first have to calculate the (unstandardized) partial slopes.

$$b_1 = \left(\frac{s_y}{s_1}\right)\left(\frac{r_{y1} - r_{y2}r_{12}}{1-r_{12}^2}\right) = \left(\frac{2.9}{1.5}\right)\left(\frac{(-0.47) - (0.56)(0.05)}{1-(0.05)^2}\right) = -2.1$$

$$b_2 = \left(\frac{s_y}{s_2}\right)\left(\frac{r_{y2} - r_{y1}r_{12}}{1-r_{12}^2}\right) = \left(\frac{2.9}{18.3}\right)\left(\frac{(0.56) - (-0.47)(0.05)}{1-(0.05)^2}\right) = 0.09$$

$$b_1^* = b_1\left(\frac{s_1}{s_y}\right) = (-0.97)\left(\frac{1.5}{2.9}\right) = -0.50$$

$$b_2^* = b_2 \left( \frac{s_2}{s_y} \right) = (0.09) \left( \frac{18.3}{2.9} \right) = 0.57$$

Cosmopolitanism appears to have the strongest effect on knowledge of family planning.

11. In order to determine the amount of variance in the dependent variable that is explained by the independent variable, we have to calculate $R^2$.

$$r_{y2 \bullet 1} = \frac{r_{y2} - (r_{y1})(r_{12})}{\sqrt{1 - r_{y1}^2} \sqrt{1 - r_{12}^2}} = \frac{(0.56) - (-0.47)(0.05)}{\sqrt{1 - (-0.47)^2} \sqrt{1 - (0.05)^2}} = 0.66$$

$$R^2 = r_{y1}^2 + r_{y2 \bullet 1}^2 (1 - r_{y1}^2) = (-0.47)^2 + (0.66)^2 [1 - (-0.47)^2] = 0.56$$

Therefore, 56% of the variance in family planning knowledge is explained by TV viewing and cosmopolitanism.

12. We can compare the bivariate correlation coefficients, but a more thorough analysis of the relationships requires that we calculate the standardized partial slopes.

$$b_1 = \left( \frac{s_y}{s_1} \right) \left( \frac{r_{y1} - r_{y2} r_{12}}{1 - r_{12}^2} \right) = \left( \frac{1.6}{1.3} \right) \left( \frac{(-0.59) - (-0.37)(0.09)}{1 - (0.09)^2} \right) = -0.69$$

$$b_2 = \left( \frac{s_y}{s_2} \right) \left( \frac{r_{y2} - r_{y1} r_{12}}{1 - r_{12}^2} \right) = \left( \frac{1.6}{23} \right) \left( \frac{(-0.37) - (-0.59)(0.09)}{1 - (0.09)^2} \right) = -0.02$$

$$b_1^* = b_1 \left( \frac{s_1}{s_y} \right) = (-0.69) \left( \frac{1.3}{1.6} \right) = -0.56$$

$$b_2^* = b_2 \left( \frac{s_2}{s_y} \right) = (-0.02) \left( \frac{23}{1.6} \right) = -0.29$$

$$r_{y2 \bullet 1} = \frac{r_{y2} - (r_{y1})(r_{12})}{\sqrt{1 - r_{y1}^2} \sqrt{1 - r_{12}^2}} = \frac{(-0.37) - (-0.59)(0.09)}{\sqrt{1 - (-0.59)^2} \sqrt{1 - (0.09)^2}} = -0.39$$

$$R^2 = r_{y1}^2 + r_{y2 \bullet 1}^2 (1 - r_{y1}^2) = (-0.59)^2 + (-0.39)^2 [1 - (-0.59)^2] = 0.45$$

There is a strong negative association between hours of study per week and number of dates (r=-0.59), and a weak-moderate negative correlation between number of dates per week and cleanliness (r=-0.37). Hours spent studying per week seems to have the strongest effect on number of dates ($b^*_1$=-0.56), with 45% of the variance in number of dates explained by these two variables. Unfortunately, that leaves an additional 55% of the variance unexplained. Keep trying, guys!

13. In order to determine which variable has the strongest effect on the dependent variable we have to calculate that standardized partial slopes. But, to calculate the standardized partial slopes, we first have to calculate the (unstandardized) partial slopes.

$$b_1 = \left(\frac{s_y}{s_1}\right)\left(\frac{r_{y1} - r_{y2}r_{12}}{1 - r_{12}^2}\right) = \left(\frac{2.3}{23}\right)\left(\frac{(0.68) - (-0.22)(-0.35)}{1 - (-0.35)^2}\right) = 0.07$$

$$b_2 = \left(\frac{s_y}{s_2}\right)\left(\frac{r_{y2} - r_{y1}r_{12}}{1 - r_{12}^2}\right) = \left(\frac{2.3}{24.3}\right)\left(\frac{(-0.22) - (0.68)(-0.35)}{1 - (-0.35)^2}\right) = 0.002$$

$$b^*_1 = b_1\left(\frac{s_1}{s_y}\right) = (0.07)\left(\frac{23}{2.3}\right) = 0.70$$

$$b^*_2 = b_2\left(\frac{s_2}{s_y}\right) = (0.002)\left(\frac{24.3}{2.3}\right) = 0.02$$

Truancy has the strongest effect on the criminal activity variable for teenagers ($b^*_1$=0.70).

14. In order to determine the amount of variance in the dependent variable that is explained by the independent variable, we have to calculate $R^2$.

$$r_{y2\bullet1} = \frac{r_{y2} - (r_{y1})(r_{12})}{\sqrt{1 - r_{y1}^2}\sqrt{1 - r_{12}^2}} = \frac{(-0.22) - (0.68)(-0.35)}{\sqrt{1 - (0.68)^2}\sqrt{1 - (-0.35)^2}} = 0.03$$

$$R^2 = r_{y1}^2 + r_{y2\bullet1}^2(1 - r_{y1}^2) = (0.68)^2 + (0.03)^2[1 - (0.68)^2] = 0.46$$

Therefore, 46% of the variance in criminal acts is explained by truancy and SES.

**SPSS or MicroCase WORK PROBLEMS**

If you had problems with any of these questions, be sure to re-read the appropriate sections at the end of the chapter in your textbook.

1.  When assessing the effects of age and siblings on the number of children, we begin by looking at the correlations between the variables. Here we see that age has a moderate positive relationship with the number of children (r = 0.399) and that siblings has a weak positive relationship with the number of children (r = 0.219). Thus as one gets older, he/she is likely to have more children, and the larger the family of origin, more children one has.

**Correlations**

|  |  | SIBS | CHILDS | REALINC | AGE | TVHOURS | SEI |
|---|---|---|---|---|---|---|---|
| SIBS | Pearson r | 1 | .219(**) | -.142(**) | .095(**) | .138(**) | -.246(**) |
|  | Sig (2-tailed) | . | .000 | .000 | .003 | .000 | .000 |
|  | N | 1002 | 1002 | 855 | 1000 | 672 | 955 |
| CHILDS | Pearson r | .219(**) | 1 | -.005 | .399(**) | .071 | -.158(**) |
|  | Sig (2-tailed) | .000 | . | .864 | .000 | .065 | .000 |
|  | N | 1002 | 1497 | 1281 | 1490 | 672 | 1425 |
| REALINC | Pearson r | -.142(**) | -.005 | 1 | -.012 | -.255(**) | .394(**) |
|  | Sig (2-tailed) | .000 | .864 | . | .655 | .000 | .000 |
|  | N | 855 | 1281 | 1282 | 1278 | 573 | 1232 |
| AGE | Pearson r | .095(**) | .399(**) | -.012 | 1 | .128(**) | .010 |
|  | Sig (2-tailed) | .003 | .000 | .655 | . | .001 | .716 |
|  | N | 1000 | 1490 | 1278 | 1493 | 672 | 1420 |
| TVHOURS | Pearson r | .138(**) | .071 | -.255(**) | .128(**) | 1 | -.246(**) |
|  | Sig (2-tailed) | .000 | .065 | .000 | .001 | . | .000 |
|  | N | 672 | 672 | 573 | 672 | 673 | 641 |
| SEI | Pearson r | -.246(**) | -.158(**) | .394(**) | .010 | -.246(**) | 1 |
|  | Sig (2-tailed) | .000 | .000 | .000 | .716 | .000 | . |
|  | N | 955 | 1425 | 1232 | 1420 | 641 | 1426 |

** Correlation is significant at the 0.01 level (2-tailed).

In the output boxes below, we see that these two independent variables explain 18.8% of the variance in number of children one has ($R^2 = 0.188$). The second output box tells us that age has a stronger effect on number of children (b* = 0.376) compared to siblings (b* = 0.183). Note that the standardized slopes are in the same direction as the relationships between each of the independent variables and dependent variable.

**Model Summary**

| Model | R | R Square | Adjusted R Square | Std. Error of the Estimate |
|---|---|---|---|---|
| 1 | .434 | .188 | .187 | 1.473 |

**Coefficients(a)**

| Model | | Unstandardized Coefficients | | Standardized Coefficients | t | Sig. |
|---|---|---|---|---|---|---|
| | | B | Std. Error | Beta | | |
| 1 | (Constant) | -.182 | .144 | | -1.268 | .205 |
| | SIBS | .097 | .015 | .183 | 6.385 | .000 |
| | AGE | .035 | .003 | .376 | 13.132 | .000 |

2. There is a weak positive correlation between age and tvhours (r = 0.128), a weak positive correlation between the number of children and TV hours (r = 0.071), a weak negative correlation between prestige and TV hours (r = -0.246), and a weak negative correlation between income and TV hours (r = -0.255). Therefore, television viewing increases with age and the number of children and decreases with prestige and income.

Looking at the regression output (box 1 below), we see that the coefficient of multiple determination ($R^2$) is 0.096, indicating that 9.6% of the variance in television hours watched per day is explained by these four variables.

The second output box below gives the regression output, including the Y intercept and standardized slopes. We can assess the effects of each of the independent variables on the dependent variable by looking at the standardized slopes. The largest of the four, sei (b*= -0.176) indicates that prestige has a stronger (though not by much) effect on TV hours watched per day than do age (b*= 0.123) and income (b*= -0.169). The number of children has a very weak effect, compared to the other independent variables (b*= -0.025).

**Model Summary**

| Model | R | R Square | Adjusted R Square | Std. Error of the Estimate |
|---|---|---|---|---|
| 1 | .310 | .096 | .089 | 2.148 |

**Coefficients(a)**

| Model | | Unstandardized Coefficients | | Standardized Coefficients | t | Sig. |
|---|---|---|---|---|---|---|
| | | B | Std. Error | Beta | | |
| 1 | (Constant) | 3.634 | .353 | | 10.296 | .000 |
| | AGE | .016 | .006 | .123 | 2.736 | .006 |
| | REALINC | .000 | .000 | -.169 | -3.784 | .000 |
| | SEI | -.020 | .005 | -.176 | -3.897 | .000 |
| | CHILDS | -.034 | .062 | -.025 | -.543 | .587 |

3.  There is a moderate positive correlation between prestige and income (r = 0.010), a weak negative correlation between the age and income (r = -0.012), and a very weak negative correlation between the number of children and income (r = -0.005). Therefore, income increases with prestige and decreases with age and children.

Looking at the regression output (box 1 below), we see that the coefficient of multiple determination ($R^2$) is 0.163, indicating that 16.3% of the variance in income is explained by these three variables.

The second output box below gives the regression output, including the Y intercept and standardized slopes. We can assess the effects of each of the independent variables on the dependent variable by looking at the standardized slopes. The largest of the three, sei (b*= 0.410) indicates that prestige has a stronger effect on income than do age (b*= -0.052) and childs (b*= 0.088).

**Model Summary**

| Model | R | R Square | Adjusted R Square | Std. Error of the Estimate |
|---|---|---|---|---|
| 1 | .404 | .163 | .161 | 29180.310 |

**Coefficients(a)**

| Model | Unstandardized Coefficients | | Standardized Coefficients | t | Sig. |
|---|---|---|---|---|---|
| | B | Std. Error | Beta | | |
| 1 (Constant) | 1697.048 | 3281.708 | | .517 | .605 |
| AGE | -100.111 | 54.743 | -.052 | -1.829 | .068 |
| SEI | 663.096 | 42.927 | .410 | 15.447 | .000 |
| CHILDS | 1717.275 | 562.963 | .088 | 3.050 | .002 |

4.  There is a moderate positive correlation between age and prestige (r = 0.394), a weak negative correlation between the siblings and prestige (r = -0.246), and a weak negative correlation between TV viewership and prestige (r = -0.246). Therefore, prestige increases with age and decreases with siblings and TV viewership.

Looking at the regression output (box 1 below), we see that the coefficient of multiple determination ($R^2$) is 0.103, indicating that 10.3% of the variance in prestige is explained by these three variables.

The second output box below gives the regression output, including the Y intercept and standardized slopes. We can assess the effects of each of the independent variables on the dependent variable by looking at the standardized slopes. The largest of the three, TV viewership (b*= -.228) indicates that TV viewership has a stronger effect on prestige than do age (b*= 0.046) and siblings (b*= -0.204).

**Model Summary**

| Model | R | R Square | Adjusted R Square | Std. Error of the Estimate |
|---|---|---|---|---|
| 1 | .320 | .103 | .098 | 18.6220 |

**Coefficients(a)**

| Model | | Unstandardized Coefficients | | Standardized Coefficients | t | Sig. |
|---|---|---|---|---|---|---|
| | | B | Std. Error | Beta | | |
| 1 | (Constant) | 57.982 | 2.381 | | 24.352 | .000 |
| | AGE | .054 | .044 | .046 | 1.224 | .221 |
| | SIBS | -1.276 | .237 | -.204 | -5.380 | .000 |
| | TVHOURS | -1.913 | .318 | -.228 | -6.007 | .000 |

357